Trigonometry: A Very Short Introduction

VERY SHORT INTRODUCTIONS are for anyone wanting a stimulating and accessible way into a new subject. They are written by experts, and have been translated into more than 45 different languages.

The series began in 1995, and now covers a wide variety of topics in every discipline. The VSI library currently contains over 600 volumes—a Very Short Introduction to everything from Psychology and Philosophy of Science to American History and Relativity—and continues to grow in every subject area.

Very Short Introductions available now:

ABOLITIONISM Richard S. Newman
THE ABRAHAMIC RELIGIONS Charles L. Cohen
ACCOUNTING Christopher Nobes
ADAM SMITH Christopher J. Berry
ADOLESCENCE Peter K. Smith
ADVERTISING Winston Fletcher
AESTHETICS Bence Nanay
AFRICAN AMERICAN RELIGION Eddie S. Glaude Jr
AFRICAN HISTORY John Parker and Richard Rathbone
AFRICAN POLITICS Ian Taylor
AFRICAN RELIGIONS Jacob K. Olupona
AGEING Nancy A. Pachana
AGNOSTICISM Robin Le Poidevin
AGRICULTURE Paul Brassley and Richard Soffe
ALEXANDER THE GREAT Hugh Bowden
ALGEBRA Peter M. Higgins
AMERICAN CULTURAL HISTORY Eric Avila
AMERICAN FOREIGN RELATIONS Andrew Preston
AMERICAN HISTORY Paul S. Boyer
AMERICAN IMMIGRATION David A. Gerber
AMERICAN LEGAL HISTORY G. Edward White
AMERICAN NAVAL HISTORY Craig L. Symonds
AMERICAN POLITICAL HISTORY Donald Critchlow
AMERICAN POLITICAL PARTIES AND ELECTIONS L. Sandy Maisel
AMERICAN POLITICS Richard M. Valelly
THE AMERICAN PRESIDENCY Charles O. Jones
THE AMERICAN REVOLUTION Robert J. Allison
AMERICAN SLAVERY Heather Andrea Williams
THE AMERICAN WEST Stephen Aron
AMERICAN WOMEN'S HISTORY Susan Ware
ANAESTHESIA Aidan O'Donnell
ANALYTIC PHILOSOPHY Michael Beaney
ANARCHISM Colin Ward
ANCIENT ASSYRIA Karen Radner
ANCIENT EGYPT Ian Shaw
ANCIENT EGYPTIAN ART AND ARCHITECTURE Christina Riggs
ANCIENT GREECE Paul Cartledge
THE ANCIENT NEAR EAST Amanda H. Podany
ANCIENT PHILOSOPHY Julia Annas
ANCIENT WARFARE Harry Sidebottom
ANGELS David Albert Jones
ANGLICANISM Mark Chapman
THE ANGLO-SAXON AGE John Blair
ANIMAL BEHAVIOUR Tristram D. Wyatt
THE ANIMAL KINGDOM Peter Holland

Glen Van Brummelen

TRIGONOMETRY

A Very Short Introduction

OXFORD
UNIVERSITY PRESS

Great Clarendon Street, Oxford, OX2 6DP,
United Kingdom

Oxford University Press is a department of the University of Oxford.
It furthers the University's objective of excellence in research, scholarship,
and education by publishing worldwide. Oxford is a registered trade mark of
Oxford University Press in the UK and in certain other countries

First edition published in 2020

Published in the United States of America by Oxford University Press
198 Madison Avenue, New York, NY 10016, United States of America

British Library Cataloguing in Publication Data
Data available

Library of Congress Control Number: 2019949809

ISBN 978–0–19–881431–3

Printed and bound by
CPI Group (UK) Ltd, Croydon, CR0 4YY

Contents

Preface

This *Very Short Introduction* is not a refresher of school trigonometry, although it will cover some of that material. Rather, it aims to reveal the richness of the *entire* subject of trigonometry: its tangled history, its uses in a wide variety of scientific and practical applications, and its interactions with some of the most intriguing mathematics in existence, such as infinity, complex numbers, and non-Euclidean geometry.

You will find here an abundance of creative logical and empirical thinking, in a place where geometry and measurement come together. We will be exploring not just *what* is true, but also *why* it is true. In mathematics this often takes the form of logical arguments. We won't be formal, but we also won't avoid them. The *why* question is where mathematical creativity is born. Now, it happens to everyone (most of all, the author) that while reading a proof, at a certain point one gets stuck. This can be very frustrating. In these situations you can do two things. First, seek out help in other resources, such as the Further Reading or the Internet. Second, feel free to skip ahead to the result and pick up the thread from there. Most of the topics are modular, and you will not lose much if you abandon an argument that is causing you trouble.

Writing a *Very Short Introduction* also grants me an excuse to bypass some of the fine details. Our coverage is not intended to be thorough; that is not the point of this series. Other resources (again, see Further Reading) should help to answer questions that fall between the cracks.

Finally, this book is accessible to anyone with some understanding of mathematics. That is not to imply that you will necessarily understand everything perfectly. It does mean that you will be rewarded with an inkling of the magic and mystery of mathematics in general, and trigonometry in particular. An understanding of calculus is not assumed anywhere (even in Chapter 5 on infinity), except for the odd moment here and there where I state a formula or two to illustrate a point for those familiar with the subject.

Enter this book prepared to engage your sense of wonder. I did, and it's been a remarkable experience.

Acknowledgements

I have been blessed with the kindness of many people who have supported me by sacrificing some of their free time to critique and improve this book. First on that list are my students at Quest University and at the summer camp Mathpath; it is to them that I dedicate this book. Several readers deserve special mention: Adam Achs, Ferdinand Gruenenwald, Clemency Montelle, Kailyn Pritchard, Michele Roblin, Ariel Van Brummelen, and Venessa Wallsten. The team at Oxford University Press (Latha Menon, Jenny Nugée, Sandy Garel and others whose names regretfully I likely will never know) were outstanding. They are the reason why the *Very Short Introductions* series has been of such high quality and enjoyed such success. Finally, my thanks go to the reviewers of the original proposal and of the manuscript. Their efforts surely have decreased substantially the need for me to apologize for deficiencies. Nevertheless, I still do so.

List of illustrations

Chapter 1
Why?

Hipparchus

Hipparchus of Rhodes had a problem. The ancient Greek astronomer was trying to predict the times of eclipses—those dramatic moments in the heavens when the Moon, Sun, and Earth come into perfect alignment, blocking from view either the Sun or the Moon. Knowing when they were going to happen would bring a powerful sense of understanding and mastery of the universe. But it's difficult to do. The Moon moves very quickly through the backdrop of the fixed stars (more than 10° per day, twenty times its diameter), and its path is irregular. The Sun also moves, although more slowly and predictably: about 1° per day, along a perfect circle known as the *ecliptic*. Yes, the Sun is travelling around the Earth: we're being *ancient* astronomers here, not *modern*.

Let's look at that path more closely (Figure 1). The Sun takes exactly one year to traverse the entire circle; after all, that's what a 'year' means. However, Hipparchus knew that the Sun does not travel at a constant speed along the circle. In ancient times it travelled most slowly during the spring. As hard as it is to believe typing these words in September, at present the Sun moves most slowly during the summer, making summer the longest season. The ancient mathematical toolset was not designed to work

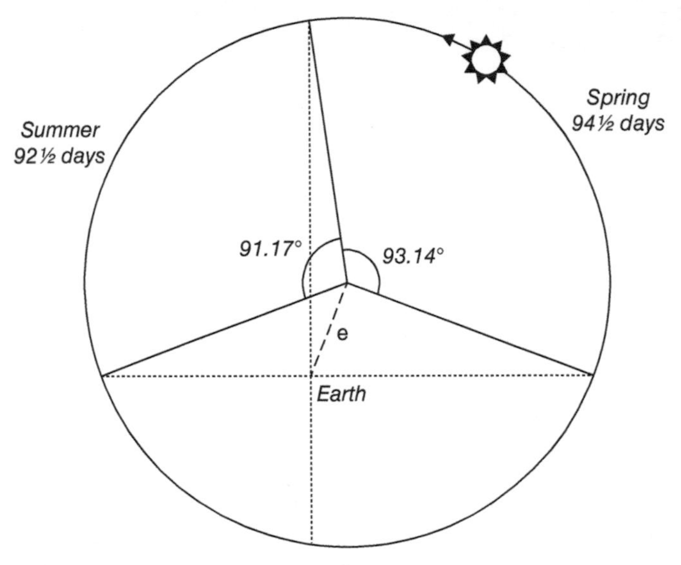

1. **Hipparchus of Rhodes' solar model.**

directly with objects that speed up and slow down on their circular paths. Rather, Hipparchus moved the Earth away from the centre of the Sun's orbit. In this way the Sun could travel at a constant speed on the ecliptic; but when it is further away from us in the spring, it would appear to us to be travelling more slowly.

To predict eclipses accurately, Hipparchus would need to know precisely where the Sun and Moon are at any instant. For a start, this requires knowing the Earth's distance from the centre of the Sun's orbit—its *eccentricity*, *e*. This in turn implies that we need to know the radius of the orbit. But since the ancients were unable to measure the distance to the Sun, we shall assume simply that the radius is one very large unit of distance away from us: in fact, an *astronomical unit*. Hipparchus knew the lengths of the seasons: 94½ days for the spring and 92½ days for the summer. Since a circle has 360°, we can convert these values easily to angles:

for the spring,

$$(94\tfrac{1}{2}\text{ days})\cdot\frac{360^\circ}{365\frac{1}{4}\text{ days}} = 93.14^\circ;\text{ and for the summer,}$$

$$(92\tfrac{1}{2}\text{ days})\cdot\frac{360^\circ}{365\frac{1}{4}\text{ days}} = 91.17^\circ.$$

But now Hipparchus was stuck. We have measures of arcs and angles in Figure 1, but other than the radius of the circle (which we picked out of thin air in any case), we don't have any information about lengths. Nor do we have any tools at hand that allow us to find them. If Hipparchus couldn't convert angles to lengths, he wasn't going to predict any eclipses. He wasn't even going to be able to take the first step and determine the value of e.

Bressieu

Maurice Bressieu had a problem. The 16th-century French mathematician and humanist was writing a treatise on mathematics and astronomy, eventually published as *Metrices astronomicae* (*Astronomical Measurement*) in 1581. Somehow, he became interested in a different problem: how to determine the height of a nearby tower (Figure 2). Until now his mathematical work had supported astronomical endeavours in the heavens, but this problem was entirely earthly. He measured fifty paces along the ground from a vantage point C to the base of the tower B; and using an instrument to measure angles, he found that the angle BCA was 60.5°. But where to go now?

Bressieu could use techniques from 'practical geometry', a medieval practice that used simpler tools than mathematical astronomy. For instance, he could make a smaller triangle with the same angle at C and measure the lengths of its sides. Since the two triangles are the same shape, he could use these results to find the tower's height. But there had to be a way to use the more elegant

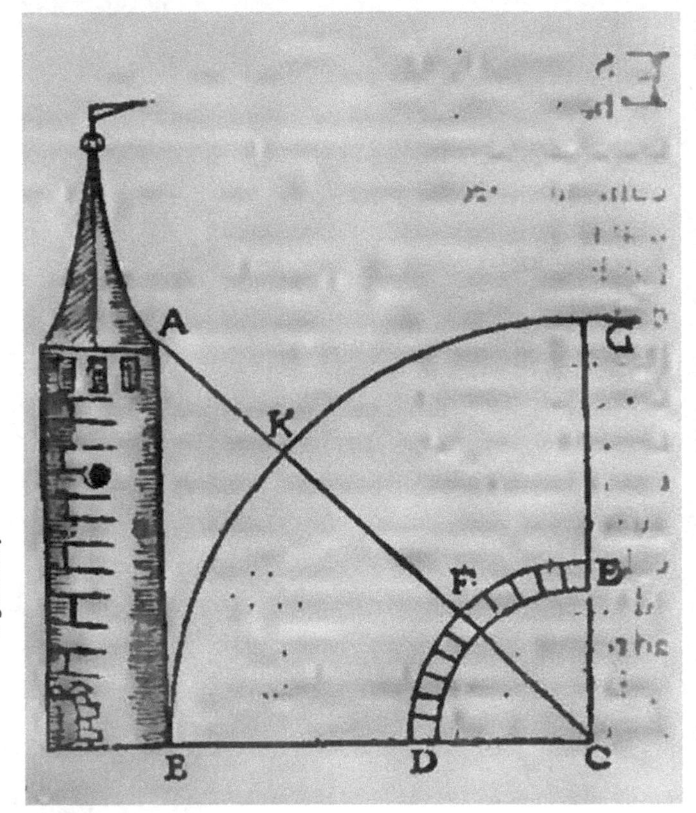

2. Finding the altitude of a tower in Maurice Bressieu's *Metrices astronomicae* (1581).

and powerful mathematics he had just been writing about, even if it meant bringing the mathematics of the heavens down to the mundane world of towers and paces.

Lord Kelvin

Sir William Thomson had a problem. The physicist who dominated Victorian science and would later become Lord Kelvin

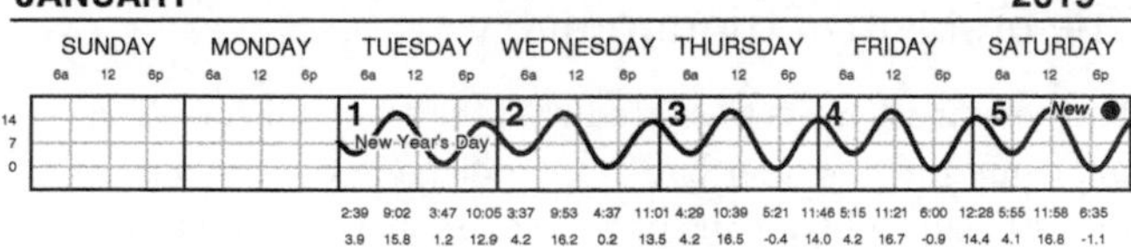

3. A graph of ocean tides for Ketchikan, Alaska, over five days.

had become interested in predicting the behaviour of ocean tides. Knowing the state of a tide in advance can be immensely valuable to sailors, fishermen, coastal engineers, and surfers. It could even be a matter of life and death: the ships involved in Julius Caesar's first landing in Britain in 55 BC were destroyed because they were unaware of the local tides. The rise and fall of a tide occurs in regular patterns (Figure 3), but simply looking at a graph of the tides over the past several days does not provide enough information for us to judge reliably how it might continue over the next few days, let alone months.

The astronomical influences on tides were well known by Thomson's time, namely, gravitational interactions between the Moon, the Sun, and the Earth. Our ability to predict their relative positions had advanced astronomically beyond what Hipparchus had accomplished two millennia earlier. But there is much more to the story: various local effects such as the slope of the harbour or the water's depth can make a significant impact on how a tide will behave from one place to the next. Thomson's problem then boiled down to this: we have a chart of tidal observations at a certain place, say, over the past month. From this chart alone, we need to separate the various influences that led to this tidal pattern, and then combine our knowledge of these influences to predict the same tide going forward. This is no easy task. Physicist Richard Feynman once remarked about a similar situation: 'It is easy to make a cake from a recipe; but can we write down the recipe if we are given a cake?'

The problem of trigonometry

Each of our scientists was faced with the same difficulty. The mathematical tools they had at their disposal were not up to the task of quantifying the phenomenon they were studying. Hipparchus could provide accurate measurements of the arcs corresponding to the seasons, but he could not transfer this information to the determination of the lengths of lines within the same diagram. Bressieu could see from his diagram of the tower that he had enough information to find its height in principle; there could be only one tower fifty paces away that corresponded to an altitude of 60.5°. But again, how could he convert knowledge of this angle into knowledge of a distance?

This is the problem at the heart of trigonometry: how can we bring together geometry and computation to solve real physical problems? In a way, this is the origin of science. If Hipparchus could convert his geometric hypothesis concerning the Sun's motion into predictions of its future locations, he would be in a position to conduct a genuine scientific experiment. Today, examples of bridges between geometry and measurement are everywhere. A software designer wants to rotate an image on a computer screen as part of an animation for the latest video game. A welder needs to cut a steel beam into the right length to fit between two curved surfaces. A surveyor needs to know the area of an irregularly shaped tract of land. If mathematics is the discipline of both geometry and number, trigonometry is what allows us to travel back and forth between them.

You may have noticed that I left Kelvin's tide predictions out of this characterization of trigonometry. There is something qualitatively different about this example, and that difference reflects a dramatic change in the nature of the subject that occurred during the 18th century. For now, I will leave this omission as a curiosity to ponder; we will revisit the matter in Chapter 5.

Chapter 2
Sines, cosines, and their relatives

Let's return to the height of Maurice Bressieu's tower in Figure 2. The path from the observer's location C, to the base B of the castle, to the top A, and back to C forms a right triangle. The angle at the observer is $60.5°$, so the angle at the top must be $29.5°$, so we know the triangle's shape. We also know that the base is fifty paces long. How to find the lengths of the other two sides? Before Bressieu this topic was the domain of practical geometry, and this unassuming name was well chosen. Hugh of St Victor, a 12th-century mystical theologian who took time out from his heavenly ponderings to write on the subject, simply used an astrolabe (a circular plate used to measure the altitudes of stars) to construct a small triangle identical in shape to the large one (i.e. *similar* to it; see Figure 4). The base of this triangle on the astrolabe was traditionally broken into twelve units. Using a grid on the astrolabe, he could measure the height of the smaller triangle; it comes out to a little more than twenty-one units. Since the ratio of the height to the base is $\frac{21}{12}$, the height of the tower must be roughly $\frac{21}{12} \cdot 50 = 87.5$ paces. (As one can see from the diagram, an angle greater than $45°$ at C makes the measurement harder, since the triangle extends upward past the edge of the astrolabe. Hugh likely would have advised Bressieu to back up a bit.)

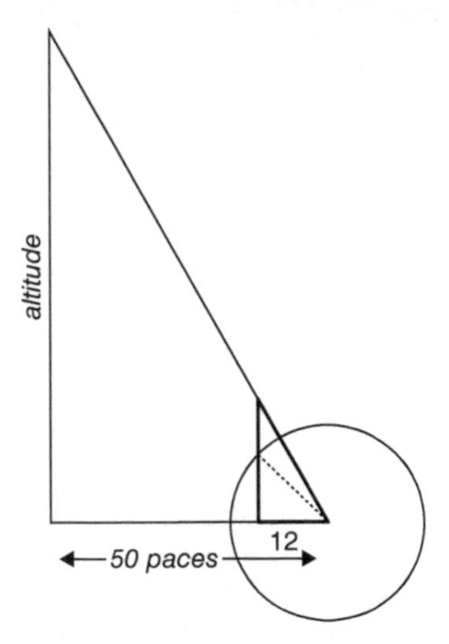

4. Hugh of St Victor finding the height of Bressieu's tower.

Meanwhile, Bressieu had access to much more precise mathematics for use in his astronomy, a discipline that his predecessor Regiomontanus had called 'the foot of the ladder to the stars'. Bressieu was nervous about how his colleagues might react to bringing this celestial subject down to the Earth; he confined his work on the tower to an appendix and began it with the phrase, 'We hope it will not be unpleasant to the reader . . .'. This subject, which was then known as the 'science' or the 'doctrine' of triangles, would receive its final appellation fourteen years later in Bartholomew Pitiscus' 1595 book *Trigonometriae*, the 'measurement of triangles'. As one can see from the subtitle on his cover (Figure 5), by then scholars had gotten over their skittishness about using trigonometry for practical applications.

Defining and using the basic trigonometric functions

To find the height of Bressieu's tower, we will need to find from the given angle the ratios of the sides to each other, just as from

Bartholomæi
Pitisci Grunbergensis
Silesij:
TRIGONOMETRIÆ
Siue;
De dimensione Triangulor:
LIBRI QVINQVE. Item
PROBLEMATVM VARIORV.
nempe
Geodæticorum,
Altimetricorum,
Geographicorum,
Gnomonicorum, et
Astronomicorum:
LIBRI DECEM.
TRIGONOMETRIÆ SVBIVNCTI,
AD VSVM EIVS DEMON-
STRANDVM.

Sancti Vrbani Bibliothecæ,

Sumbtib. et impens.
Dominci. Custodis
chalcographi.
M. DC..

5. **The title page of the 1600 edition of Pitiscus's *Trigonometriae*, the first appearance of the word 'trigonometry'.**

the angle 60.5° we found the ratio $\frac{21}{12}$ between the altitude and the base. We call the known angle θ, and we'll name the sides the *Hypotenuse*, the side *Opposite* the known angle, and the side *Adjacent* to the known angle (Figure 6). Next, we give names to the various ratios between the sides:

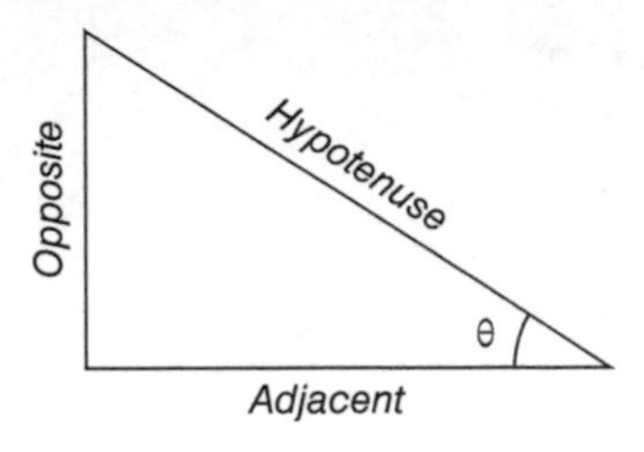

6. The names of the three sides in a right triangle.

$$\text{the } \textbf{\textit{sine}} \text{ of } \theta, \sin\theta = \frac{\text{Opposite}}{\text{Hypotenuse}};$$

$$\text{the } \textbf{\textit{cosine}} \text{ of } \theta, \cos\theta = \frac{\text{Adjacent}}{\text{Hypotenuse}};$$

$$\text{the } \textbf{\textit{tangent}} \text{ of } \theta, \tan\theta = \frac{\text{Opposite}}{\text{Adjacent}}.$$

Virtually every English-speaking student memorizes these definitions using the mnemonic SOH-CAH-TOA: '*S*ine is *O*pposite over *H*ypotenuse, *C*osine is *A*djacent over *H*ypotenuse, and *T*angent is *O*pposite over *A*djacent'. For instance, in the case of Bressieu's tower, the ratio of the Opposite side to the Adjacent side (about $\frac{21}{12}$) is called the tangent of 60.5°, or $\tan 60.5°$.

We still don't know what any of these ratios are, but any scientific calculator does. *How* it knows them is a question we'll address in Chapters 3 and 5. In the meantime, you can put your calculator to the test by entering $\tan 60.5°$ (make sure it's in *degrees* mode, not *radians*). You should get 1.767494, which is indeed quite close to $\frac{21}{12}$.

Let's try a modern example. Every day I walk home by cutting diagonally across a soccer field, a journey of 125 metres. Today my path is blocked by a new sign that says 'keep off the grass' (Figure 7). How much longer will I have to walk? I begin by measuring the angle between my diagonal journey and the

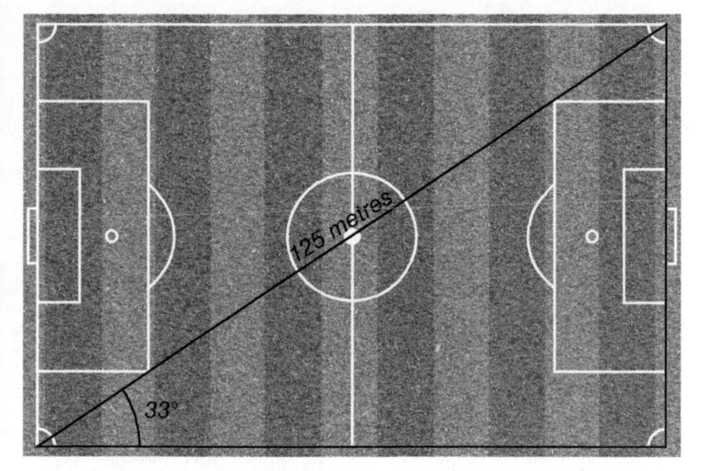

7. Walking across a soccer field.

sideline, getting a value of 33°. The width of the field is the opposite side and the diagonal route is the hypotenuse. So

$$\sin 33° = \frac{\text{Opposite}}{\text{Hypotenuse}} = \frac{\text{Width of field}}{125}, \text{ or}$$
$$\text{Width of field} = 125 \sin 33° = 68.1 \text{ metres.}$$

We may find the field's length in a similar way:

$$\cos 33° = \frac{\text{Adjacent}}{\text{Hypotenuse}} = \frac{\text{Length}}{125}, \text{ or}$$
$$\text{Length} = 125 \cos 33° = 104.8 \text{ metres.}$$

Altogether I will need to walk $68.1 + 104.8 = 172.9$ metres, almost 48 metres longer than my diagonal shortcut.

We can also get a precise value for the height of Bressieu's tower:

$$\tan 60.5° = \frac{\text{Opposite}}{\text{Adjacent}} = \frac{\text{Height}}{50 \text{ paces}}, \text{ so}$$
$$\text{Height} = 50 \tan 60.5° = 88.4 \text{ paces.}$$

Armed with these methods, we are now able to tackle more complicated problems. Recall our ancient astronomer Hipparchus, who wanted to find the eccentricity of the Sun's orbit around the Earth (Figure 1). Assuming that the circle's radius is equal to 1, our goal is to find the distance e from the Earth to the circle's centre. There are no right-angled triangles in the diagram. But if we draw horizontal and vertical lines through the centre of the circle (Figure 8), we can get started. The 93.14° angle corresponding to the spring has now been broken into a right angle and two smaller angles, which we'll call α and β. We can find α by adding the spring and summer angles together. From the figure, we see that the total of 184.31° is also equal to the upper semicircle plus two copies of α. So, $\alpha = \frac{184.31° - 180°}{2} = 2.155°$. But we can assemble the summer angle by taking the upper left right angle, adding α, and subtracting β. Thus $91.17° = 90° + \alpha - \beta$, which tells us that $\beta = 0.985°$.

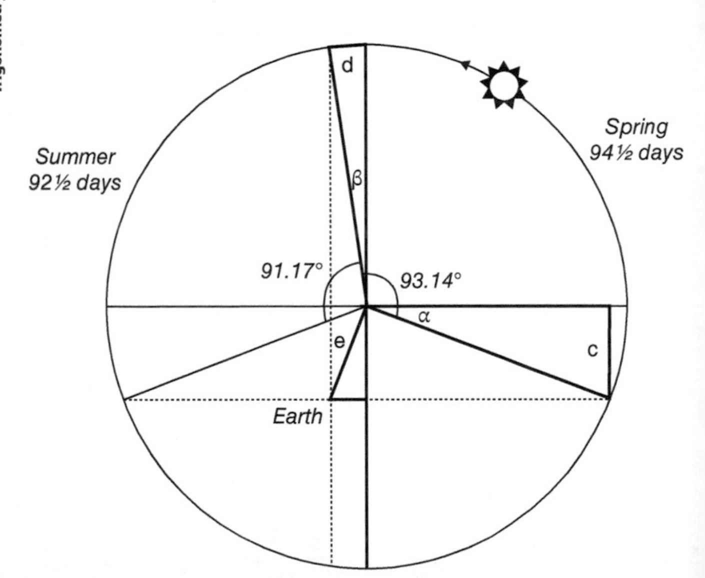

8. The solution to Hipparchus' solar eccentricity problem.

Next, let's focus our attention on the three right-angled triangles drawn with bold lines. For two of these triangles the hypotenuse is the radius of the circle, which has length 1. From the triangle on the right, we can find the opposite side c:

$$\sin 2.155° = \frac{c}{1}, \text{ so } c = 0.0376.$$

From the triangle on the top we can find the opposite side d:

$$\sin 0.985° = \frac{d}{1}, \text{ so } d = 0.0172.$$

Turning to the bold triangle in the middle, its hypotenuse is equal to e and its other two sides are c and d. So, using the Pythagorean theorem (in a right-angled triangle, the square of the hypotenuse is equal to the squares of the other two sides),

$$e = \sqrt{c^2 + d^2} = \sqrt{0.0376^2 + 0.0172^2} = 0.0414.$$

There we have it; we have solved the world's first trigonometric problem.

History is never as simple as it seems, and a couple of disclaimers are needed here. First, Hipparchus didn't have the sine function, or in fact any of our modern trigonometric functions. Rather, he had a table that allowed him to find the length of a chord within a circle (Figure 9), which is twice the sine of half the angle corresponding to the chord. This is the beginning of the long story that led to our 'sine'. In early India around the 5th century AD, astronomers came to realize that dividing Hipparchus' chord by two would simplify their calculations. In Sanskrit, they named this new length the *jya-ardha*, or half chord. Eventually the *ardha* fell away, and when the *jya* was transmitted into Arabic, it was transliterated as *jayb*. It turns out that *jayb* already had some related meanings: a bay, or cavity, or bosom. So when *jayb* found its way to Latin Europe, it was translated as *sinus* (yes, as in your

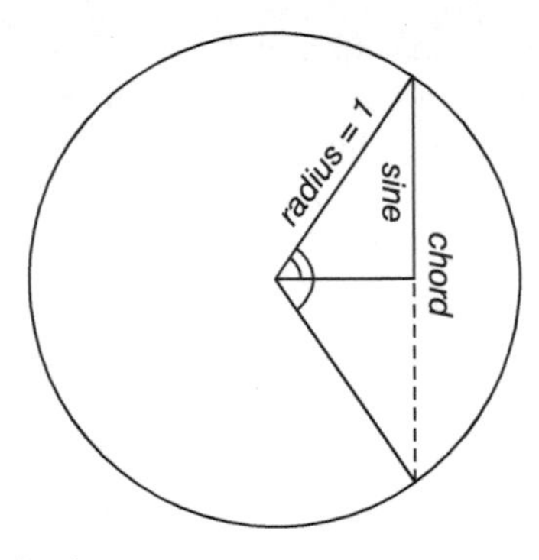

9. The chord and the sine.

sinus cavity). It still goes by this name in some countries, although in English it has been shortened to *sine*.

Our abbreviations 'sin' and 'cos' haven't always been universally accepted. Until the late 18th century it was common simply to use 'S' or 's' for 'sine', and various abbreviations including 'c', 'cs', and even 'Σ' for 'cosine'. Perhaps the most inventive symbols were those devised by late 19th-century logician Charles Dodgson (better known as Lewis Carroll): ⏜ for 'sine', and ⌓ for 'cosine'.

Second, ancient Greek astronomers did not use a circle of radius 1, our 'unit circle'. Rather, they used a circle of radius 60, deriving from the Babylonian base 60 number system they used in their astronomical work. Many different radii have been used over the centuries. A common value in India was the seemingly peculiar 3,438, which turns out to be the length of the circle's radius if you set one minute of arc ($\frac{1}{60}$ of a degree) to be your unit of length. In 15th- and 16th-century Europe radii such as 100,000 or even 10,000,000 were used. At that time our positional decimal system for fractions was still in the future; such large radii allowed astronomers to represent sines as whole numbers. The circle of radius 1 had appeared first in 10th-century Iraq, but it didn't show up in Europe until much later. One implication of this way of thinking is that the sine was considered to be the *length* of the opposite side in the triangle, not the *ratio* between the opposite side and the hypotenuse as we think of it today.

Less common functions

You might have wondered why we defined only three trigonometric functions using the ratios of the Opposite, Adjacent, and Hypotenuse sides of the right triangle. There are three other possibilities:

$$\text{the } \textit{cosecant} \text{ of } \theta, \operatorname{cosec} \theta = \frac{\text{Hypotenuse}}{\text{Opposite}};$$

$$\text{the } \textit{secant} \text{ of } \theta, \sec \theta = \frac{\text{Hypotenuse}}{\text{Adjacent}};$$

$$\text{the } \textit{cotangent} \text{ of } \theta, \cot \theta = \frac{\text{Adjacent}}{\text{Opposite}}.$$

We can solve triangle problems just as easily with these new functions. For instance, imagine a surveyor building a bridge across a river joining two gates, one of which is downstream from the other. The river is 50 metres wide, and the angle between the new bridge and the bank will be 25°. How long will the bridge have to be? From the triangle in Figure 10, we know that

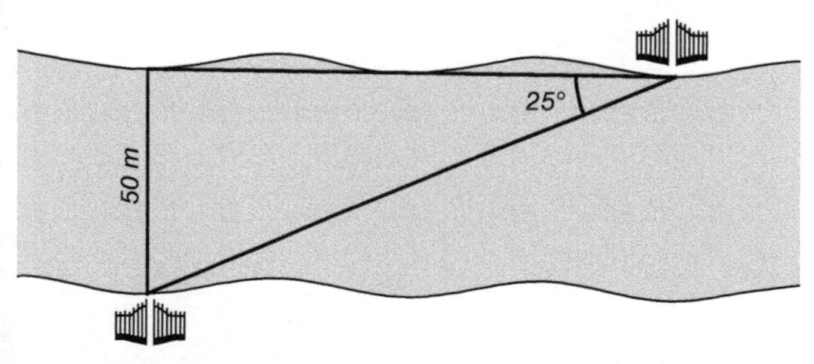

10. Building a bridge.

$$\text{cosec } 25° = \frac{\text{Hypotenuse}}{\text{Opposite}} = \frac{\text{Bridge length}}{50 \text{ metres}}; \text{so length} = 50 \text{ cosec } 25°.$$

We don't have a button on our calculator for cosec. However, we do know that cosec is the reciprocal of the sine (it's Hypotenuse/Opposite, rather than Opposite/Hypotenuse). Then

$$\text{Length} = 50 \text{ cosec } 25° = \frac{50}{\sin 25°} = 118.3 \text{ metres},$$

and the problem is solved.

This solution raises a question: if we can solve the problem by converting the cosecant to a sine, why bother with the cosecant at all? The simple answer is that we don't have to. The same is true of the secant (the reciprocal of the cosine) and the cotangent (the reciprocal of the tangent). We'll encounter situations later where the mathematics will look a little smoother if we use one of these minor functions, but there's a reason they are minor—they can always be replaced. This is why your calculator doesn't have buttons labelled cosec, sec, or cot.

Let's look at one more example. Consider the simplest form of sundial, a stick in the ground (called a gnomon) in Figure 11.

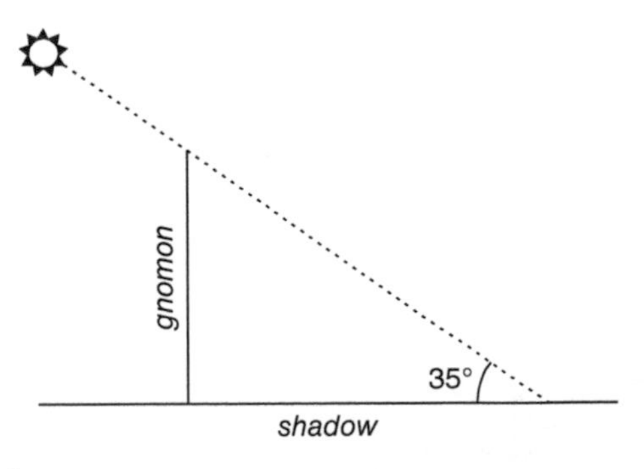

11. A sundial.

Suppose that the gnomon is 80 cm tall. If the Sun's altitude is 35°, how long is the shadow? Using the triangle formed by the gnomon and the shadow, we have

$$\cot 35^\circ = \frac{\text{Adjacent}}{\text{Opposite}} = \frac{\text{Shadow}}{\text{Gnomon}} = \frac{\text{Shadow}}{80}, \text{ so}$$
$$\text{Shadow} = 80 \cot 35^\circ = 80 / \tan 35^\circ = 114.25 \text{ cm.}$$

What's in a name?

The cotangent function is a natural choice to find the length of the shadow of a vertical sundial. For a sundial with a horizontal gnomon, for instance on the wall of a building, the tangent function is used. This is why, during the medieval period in the Islamic world and Europe, these two functions were called the *shadows* (in Latin, *umbra versa* and *umbra recta*). The origin of the English word 'tangent' is a little more involved. When 15th-century Italian astronomer Giovanni Bianchini introduced the tangent to Europe outside of the context of sundials, it was a numerical table within his collection of astronomical tables, the *Tabulae magistrales*. He thought of it not as a trigonometric function, but as an auxiliary quantity somewhere between trigonometry and astronomy; it was helpful to him in converting stellar positions from one coordinate frame on the celestial sphere to another (see Further Reading). His successor Regiomontanus called this invention the *tabula fecunda*, the 'fruitful table', in his own work. Inspired by Regiomontanus, Italian monk Francesco Maurolico devised his *tabula benefica* (the 'beneficial table') in 1557, which turns out to tabulate what we call the secant.

But those are not the names we use today. These were born a little later in 1583 in Danish scientist Thomas Fincke's *Geometria rotundi*. We can see Fincke's inspiration in a single diagram that defines all six trigonometric functions (Figure 12). This diagram, going back as far as 10th century Baghdad astronomer Abū'l-Wafā', uses the unit circle. Since the radius is 1,

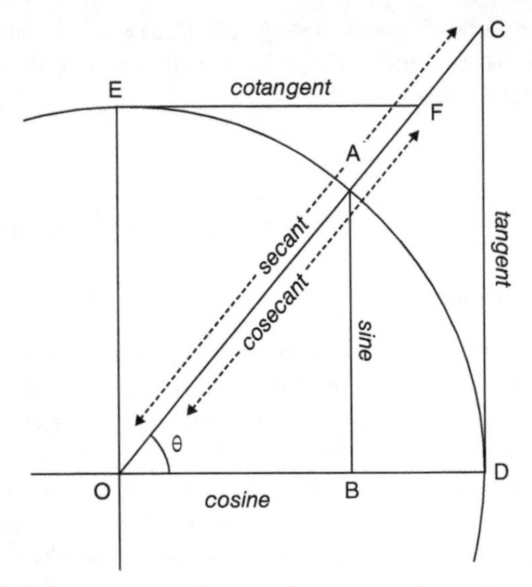

12. All six trigonometric functions in one diagram.

$$\sin\theta = \frac{\text{Opposite}}{\text{Hypotenuse}} = \frac{AB}{1} = AB\text{; similarly, } \cos\theta = OB.$$

But triangle OCD is similar to triangle OAB; hence

$$\frac{\sin\theta}{\cos\theta} = \frac{CD}{OD} = \frac{CD}{1} = CD.$$

However,

$$\tan\theta = \frac{\text{Opposite}}{\text{Adjacent}} = \frac{\text{Opposite / Hypotenuse}}{\text{Adjacent / Hypotenuse}} = \frac{\sin\theta}{\cos\theta},$$

so $CD = \tan\theta$. From the same two similar triangles we may find OC:

$$\frac{OC}{OD} = \frac{OA}{\cos\theta}\text{, but } OA = OD = 1\text{, which gives } OC = \frac{1}{\cos\theta}.$$

But since $\cos\theta = \dfrac{\text{Adjacent}}{\text{Hypotenuse}}$ and $\sec\theta = \dfrac{\text{Hypotenuse}}{\text{Adjacent}}$, $\cos\theta$ and $\sec\theta$ are reciprocals of each other. So $OC = \sec\theta$.

Within this diagram we have the reason for the names. The line CD defining the *tangent* is tangent to (i.e. touches but does not cross) the unit circle. On the other hand, the line OC defining the *secant* is secant to (cuts across) the unit circle.

We can complete the diagram by comparing triangle OEF to triangle OAB. The two angles at the circle's centre, $\measuredangle FOE$ and $\measuredangle AOB$, add up to 90°—that is, they are *complementary*. But the same is true of $\measuredangle FOE$ and $\measuredangle OFE$. Hence $\theta = \measuredangle AOB = \measuredangle OFE$, and the two triangles are similar. Thus

$$\frac{EF}{EO} = \frac{OB}{AB}, \text{ or } \frac{EF}{1} = \frac{\cos\theta}{\sin\theta}, \text{ so } EF = \frac{1}{\tan\theta} = \cot\theta;$$

and

$$\frac{OF}{EO} = \frac{OA}{AB}, \text{ or } \frac{OF}{1} = \frac{1}{\sin\theta}, \text{ so } OF = \frac{1}{\sin\theta} = \operatorname{cosec}\theta.$$

There is still a mystery in our nomenclature: why are three of the names (cosine, cosecant, and cotangent) formed by writing the prefix 'co' in front of the other three? The answer lies in the word we just defined, *complementary*. Let's return to the soccer field problem (Figure 7). We found the width to be $125\sin 33° = 68.1$ metres. We could have found it another way. Consider the angle at the top right corner of our triangle. It is complementary to the 33° angle, or $90° - 33° = 57°$. If we consider the angle at the top right to be our given angle, then

$$\frac{\text{width}}{125} = \frac{\text{Adjacent}}{\text{Hypotenuse}} = \cos 57°, \text{ so width} = 125\cos 57° = 68.1 \text{ metres}.$$

Thankfully, it's the same value that we got before. What's important here is that $\cos 57^\circ = \sin 33^\circ$, and more generally,

$$\cos\theta = \sin(90^\circ - \theta).$$

In other words, the cosine of any angle is equal to the sine of its complement. In 1620 the English mathematical instrument maker Edmund Gunter abbreviated the Latin *sinus complementi* (the 'sine of the complement'), and the word *cosine* was born.

One might anticipate the meanings of the other two terms. By similar lines of reasoning, the cosecant of an angle is the secant of its complement, and the cotangent of an angle is the tangent of its complement. One benefit of this taxonomy: if you have a table of sine values for arguments ranging from 0 to 90°, you can use it as a cosine table simply by reading it backwards from bottom to top. The same is true of secant/cosine tables, and of tangent/cotangent

2 TABVLA SINVVM, TANGENTIVM, ET SECANTIVM,

Minuta.	Sinus	Tangēs	Secans		Sinus	Tangēs	Secans	
	0	0	0	Gradus	1	1	1	
0	0	0000	10000000		174524	174550	10001524	60
1	2909	2909	10000001		177433	177459	10001574	59
2	5818	5818	10000002		180341	180369	10001626	58
3	8727	8727	10000004		183250	183279	10001679	57
4	11636	11636	10000008		186158	186189	10001733	56
5	14544	14544	10000010		189066	189100	10001788	55
6	17453	17452	10000014		191975	192010	10001844	54
7	20362	20361	10000020		194883	194920	10001900	53
8	23271	23270	10000027		197792	197830	10001957	52
9	26180	26179	10000034		200700	200740	10002015	51
10	29088	29088	10000042		203608	203650	10002074	50

13. The beginning of the table of sines, tangents, and secants in Adrianus Romanus' 1609 *Canon triangulorum sphaericorum*. The values of these functions for the first ten minutes are on the left side, and the values for 1°, 1°1′, . . . , 1°10′ are on the right. The column on the right aids users reading the tables backward for the 'co'-functions; for instance, cos 88°59′ is given as 177,433 for a radius of 10,000,000.

tables. Why, then, is the 'co-sine' the only major function with a 'co', and the secant the only minor function without one? When our six functions came into common use in the late 16th and 17th centuries, the secant was actually preferred to the cosine (see Figure 13). If calculators had been invented in the 17th century, you would have found a 'sec' button in place of our 'cos'. (It doesn't really matter whether you prefer the cosine or the secant; as we've seen, the mathematics works just as well regardless.)

Even more obscure functions

Our list of functions does not stop here. One that has fallen from grace only recently is the *versed sine*. Originally used in India along with the sine, it has a history longer than almost any other function and faded away only in the late 19th and early 20th centuries. Several of its names over the centuries reflect its striking appearance in the diagram that defines it (Figure 14): the *śara* in Sanskrit, *sahm* in Arabic, and *sagitta* in Latin all mean 'arrow'. The diagram resembles a bow and arrow with the chord as

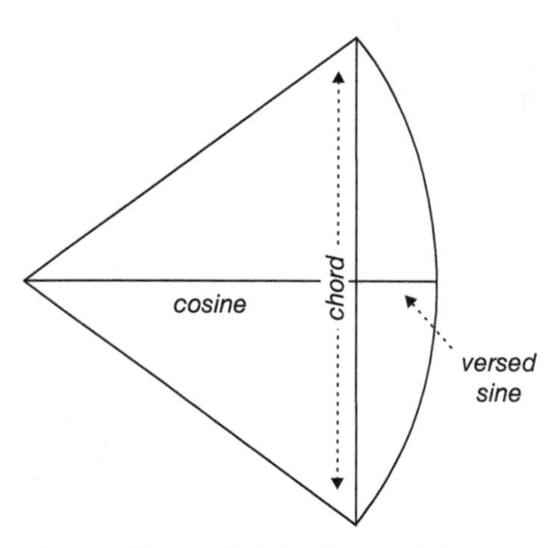

14. The bow-and-arrow diagram defining the versed sine.

the string, the cosine as the body of the arrow, and the versed sine as the arrow's tip. It is also equal to *BD* on our diagram of the trigonometric functions in Figure 12. Assuming that we are using the unit circle, the versed sine of a given angle θ is

$$\text{vers}\,\theta = 1 - \cos\theta.$$

From this definition we see that there is nothing we can do with the versed sine that we couldn't do already with the cosine. But the cosine itself is nothing more than the sine of the complement of a given angle, so all of the functions we have seen so far may be derived from the sine. Although the other functions do not give us any new mathematical power, they do make it easier to solve problems and express relationships.

The versed sine was used frequently in practical applications like astronomy, navigation, and surveying, and we'll see one in a moment. But let's begin with a purely mathematical problem: calculating π, the ratio of a circle's circumference to its diameter. In Figure 15 we have inscribed within the unit circle a regular

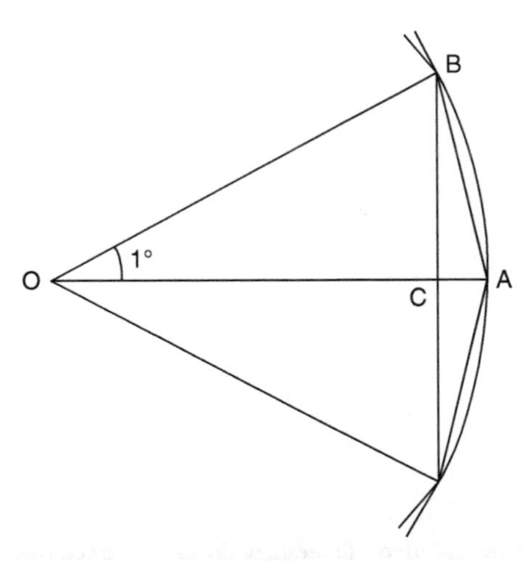

15. Estimating π with a 360-sided polygon.

polygon with 360 sides, so that each side corresponds to 1° of arc. We know that the circumference of the circle is 2π, and we hope that our 360-sided polygon is a close approximation to the circle. Now, $BC = \sin 1° = 0.0174524$ and $CA = \text{vers } 1° = 0.0001523$. Using the Pythagorean theorem, the hypotenuse of ΔABC—which is also one side of our polygon—is

$$AB = \sqrt{(\sin 1°)^2 + (\text{vers } 1°)^2} = 0.0174531.$$

Multiply by 360 to get the distance around the polygon and divide the result by 2; we get 3.1415528. Not bad.

Of course, this method requires that we compute the values of $\sin 1°$ and $\text{vers } 1°$, and if we have to rely on our calculator anyway, we might as well just use the value of π stored within it. If we want to avoid the mystery of the calculator's trigonometry buttons (which we'll clear up in Chapters 3 and 5), we could use a polygon with a better chosen number of sides. For instance, we know that the lengths of the sides of a hexagon inscribed in a unit circle are all equal to 1. But this leads to $\pi \approx 3$, a poor approximation. In the 3rd century BC Archimedes used a 96-sided polygon, and was able to find $3\frac{10}{71} < \pi < 3\frac{1}{7}$. Much later, in the 15th century, Persian astronomer Jamshīd al-Kāshī used a polygon with an astounding 805,306,368 sides to calculate π to an equivalent of sixteen decimal places. Clearly, the calculation of π is intimately related to trigonometry. We will see in Chapter 5 that the connection continued as time progressed, and better methods were found to calculate more and more digits.

The trigonometric zoo contains several other exotic creatures. The *versed cosine*, the distance in Figure 12 from B to the leftmost point of the circle, is

$$\text{vercos } \theta = 1 + \cos \theta.$$

The *exsecant*, the length AC on the secant line in Figure 12 beyond the edge of the circle, is

$$\text{exsec } \theta = \sec \theta - 1.$$

And the *excosecant*, the length AF, is

$$\text{excosec } \theta = \text{cosec } \theta - 1.$$

A sighting of any of these functions today in anything other than a playful context would be a remarkable find.

There is one reason why the versed sine and the versed cosine have a real practical benefit. It's related to the answer to another conundrum: why is it that trigonometric quantities like sines and cosines can sometimes take on negative values, even though they refer to geometric lengths and ratios? Imagine the following scenario: you are sitting on an observation platform on the ocean, watching a boat sail around you along a perfect circle 200 metres in diameter. You aren't at the centre of the circle, but rather 50 metres away from the western edge of the circle (Figure 16). The boat begins at the eastern edge of the circle and travels anticlockwise. For a given angle θ, how far eastward from you is the boat?

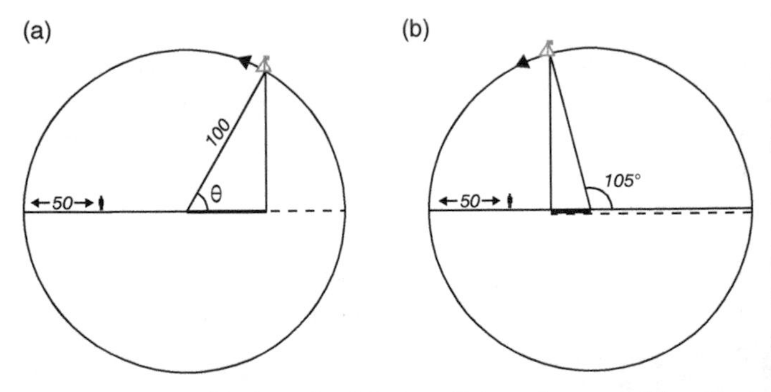

16. An example of the benefit of the versed sine: a boat travelling along a circular path.

This problem isn't very difficult if we draw a line southward from the boat to the east–west axis of the circle. Using what we've learned so far, we deduce that the distance eastward from the centre of the circle to the boat is the bolded distance in Figure 16a, $100 \cos \theta$. But we are 50 metres west of the centre, so our final expression is

$$\text{eastward distance} = 50 + 100 \cos \theta.$$

This expression works just fine for values of θ up to 90°; for instance, when $\theta = 60°$ the eastward distance is $50 + 100 \cos 60° = 100$ metres. But what if θ is greater than 90°, say, 105° (Figure 16b)? Now the boat has travelled past the northern point of the circle and is west of the circle's centre. If we follow our geometric definition of the cosine and allow $\cos 105°$ to be the positive distance on the east–west axis, then our formula will be in trouble: it tells us to add $100 \cos \theta$ (which is a *westward* distance) to the *eastward* 50 metres.

The solution is to make $\cos\theta$ a *directional* distance: in this case, eastward. Define $\cos 105°$ to be negative, and the term $100 \cos 105° = -25.88$ metres indicates that the ship is −25.88 metres east of the centre of the circle. In other words, it is 25.88 metres west of the centre. Now we can safely add it to the 50 metres, and the boat is $50 - 25.88 = 24.12$ metres east of us. Our formula now works for all values of θ.

This is all fine and well, but suppose we are at sea before the era of pocket calculators, and messing up navigational calculations is a matter of life and death. We have access to a sine/cosine table for arguments up to 90°, so we can find the length of the westward line from the centre of the circle, $100 \cos 75° = 25.88$ metres, and subtract from 50 metres. However, this procedure requires that sometimes we add the trigonometric quantity to 50 metres and sometimes we subtract, depending on the value of θ. Because we're highly competent, most of the time we'll be fine. But make a mistake just once by adding rather than subtracting, and instead of sailing on course we've put our boat somewhere on the beach.

The versed sine is a clever way to avoid this problem. Returning to Figure 16, we see that the boat's eastward distance may also be found by taking the entire distance from our location to the easternmost point of the circle, and subtracting 100 times the versed sine (the dashed line in both configurations):

$$\text{eastward distance} = 150 - 100 \text{ vers } \theta.$$

Following the boat around the circle, we see that the versed sine is always positive, so we will always subtract $100 \text{ vers } \theta$—no need to decide whether to add or subtract. This practice makes the process much more reliable, and we'll end up on the rocks less often.

We'll see in Chapter 3 that $\text{vers } \theta$, equal to $1 - \cos\theta$, is also equal to

$$2\left[\sin\left(\frac{\theta}{2}\right)\right]^2.$$

Since the sine term is squared, it is always positive. Starting in the early 19th century, navigators used a slight variation on this function—half of the versed sine, or the *haversine*—as the basis of much of their trigonometric work:

$$\text{hav } \theta = \frac{1}{2}(1 - \cos\theta) = \left[\sin\left(\frac{\theta}{2}\right)\right]^2.$$

Before moving on, a quick observation. The nested parentheses in the square of the sine term above are a bit awkward, but the square brackets are necessary. If we wrote just $\sin\left(\frac{\theta}{2}\right)^2$, this could mean either the square of $\sin\left(\frac{\theta}{2}\right)$ or the sine of $\left(\frac{\theta}{2}\right)^2$. To avoid this confusion, we move the square immediately after the 'sin':

$$\sin^2\frac{\theta}{2}.$$

This admittedly peculiar notation at least makes it crystal clear that it is the sine being squared, not the angle within it. This practice is over three centuries old, and was used by William Jones as early as 1710.

Going backwards: the inverse trigonometric functions

If trigonometry is a bridge from knowing angles to knowing lengths, then it should be possible to cross that bridge in the opposite direction. Often you have information about certain lengths and you want to find an angle. For example, suppose you need to find the angle of altitude of the Sun, an important quantity for determining the time of day or for navigation. Assuming you've carelessly left your sextant at home, you can use a nearby sundial, or just place a vertical stick in the ground (Figure 11). Our gnomon is 80 cm high and the shadow's length is 115 cm. What is the Sun's altitude?

In our triangle the opposite side is 80 cm and the adjacent side is 115 cm, so

$$\tan\theta = \frac{80}{115}.$$

From here we appear to be stuck. We could guess many different values for θ to try to find one whose tangent is close to $\frac{80}{115}$, but that strategy hardly seems efficient. Instead, our calculator has built within it a process that reverses the tangent. To take the *inverse tangent*, enter $\frac{80}{115}$ and use the button labelled '$\tan^{-1}$', or sometimes [INV] [TAN]. The calculator gives us the Sun's altitude:

$$\theta = \tan^{-1}\left(\frac{80}{115}\right) = 34.82°.$$

Again the calculator has done magic for us; again, we will explain this magic in Chapter 5.

Let's look at one last example, with both historical and modern relevance. Imagine you are on top of a mountain, or in the cockpit of an aeroplane. Due to the curvature of the Earth, if you look at the horizon you are looking slightly below the 'horizontal'. If you know how high you are above the Earth's surface, can you work out this *dip angle*? This question was considered already in the 10th century AD by Baghdad mathematician Abū Sahl al-Kūhī, and was converted to a calculation almost two centuries later by Iranian scientist Ibn Yaḥyā al-Samaw'al al-Maghribī in his *Exposure of the Errors of the Astronomers*.

Al-Samaw'al's calculation assumes the Earth's radius to be 12,982,000 cubits (about 6,000 km). For his example he takes a very small mountain indeed, $EH = 1$ cubit (around half a metre; see Figure 17). In real life, of course, we will have a much taller mountain or hopefully a much higher aeroplane. We gaze down from our lofty height to the horizon at D. Draw the radius from Z to the horizon point D, forming the right triangle ZDH. Then we can find the other side of the triangle:

$$HD = \sqrt{ZH^2 - DZ^2} = \sqrt{12\,982\,001^2 - 12\,982\,000^2} = 5.095 \text{ cubits.}$$

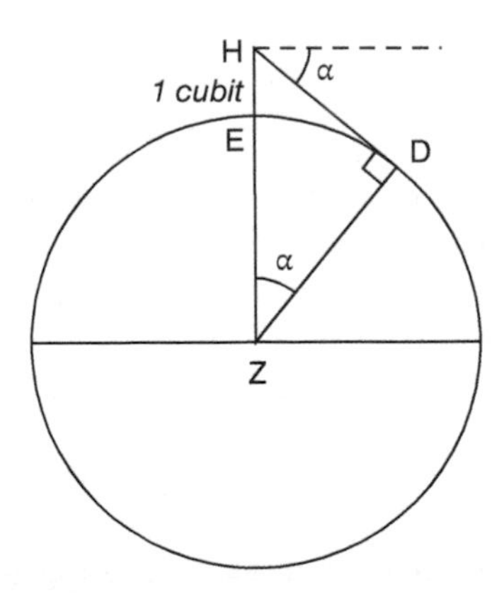

17. Finding the dip angle to the horizon.

We can use our new inverse trigonometry to find the angle at Z:

$$\measuredangle DZH = \alpha = \sin^{-1}\left(\frac{DH}{HZ}\right) = \sin^{-1}\left(\frac{5.095}{12982001}\right) = 0.0224888°.$$

But α and the dip angle are both complementary to $\measuredangle ZHD$, so α *is* the dip angle! (We've simplified the mathematics here. For the full story and an account of the 'error' that al-Samaw'al thought he was correcting, see Further Reading.)

We've already seen that our notations for working with trigonometric functions can be confusing; this trend does not end with inverse trigonometry. Recall that we chose to write $(\sin x)^2$ as $\sin^2 x$. How then should we read $\sin^{-1} x$: as the inverse sine of x, or as the totally different expression $(\sin x)^{-1}$, which means $1/\sin x$? This is a debate with a long history. English astronomer John Herschel introduced the notation '$\sin^{-1}$' for the inverse sine in 1813, consistent with the practice of writing the inverse of function f as f^{-1}. Various other notations such as $\sin^{[-1]}$ and $\overline{\sin}$ were attempted in the 19th century. The competing but clumsier notation 'arcsin' for inverse sine, meaning the arc corresponding to the given sine, does not fall prey to this confusion. Even a few decades ago 'arcsin' was still common; every once in a while you can still find a calculator that uses it. But today, the consistency of $\sin^{-1}$ with the f^{-1} notation seems to have prevailed.

Graphs of the trigonometric functions

To conclude the chapter, let's look at the trigonometric functions from a different (excuse the pun) angle. So far we've been considering the sine, cosine, and the others as geometric quantities: ratios between line segments in a triangle. But we may also consider the sine as a *function*: that is, given any value x, let $y = \sin x$. If we draw a graph of this function we get the wavy

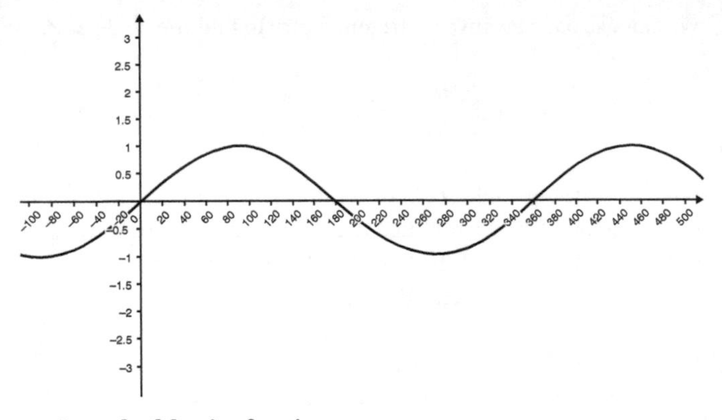

18. A graph of the sine function.

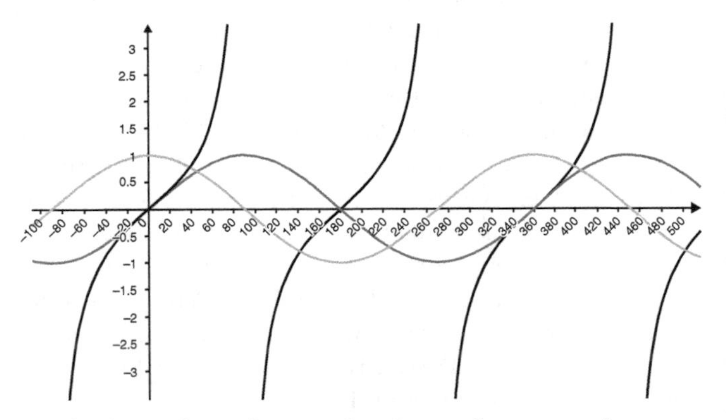

19. The sine, cosine, and tangent functions on the same graph.

pattern of Figure 18, representing the values of the sine rising and falling as the angles increase.

Figure 19 shows the graphs of the cosine and tangent functions along with the graph of the sine. Notice that the cosine graph is just the sine graph shifted to the left by 90°; in other words,

$$\cos\theta = \sin(\theta + 90°).$$

This way of thinking is very different from earlier in this chapter. By considering y to be a function of x, we almost forget that x represents an angle, and y is the ratio of sides in a right angled triangle. We might come close to saying that the sine *is* the wavy curve. *Do not think like this*; it is a dangerous misconception. The wavy curve merely represents sine values as the angles change; the sine was, is, and always will be the ratio of sides in a triangle.

Even so, there is an advantage to thinking like this. Many phenomena exhibit properties that change in periodic patterns resembling the sine wave. For instance, various mathematics textbooks discuss the population of elk on Reading Island in Canada (which, as far as I can tell, does not really exist). Let's suppose that the average elk population is 500, but fluctuates throughout the year between 450 and 550. We can use the sine wave to model this fluctuation and predict the number of elk at a given time of the year. We start with $y = \sin t$, where t is the number of months since the calving season in March. Now, the sine graph repeats its cycle every 360°, but we want the cycle to repeat every twelve months, so we alter the equation to $y = \sin\left(\frac{360}{12}t\right) = \sin(30t)$. This graph has an average value of zero, but we want the population to have an average value of 500, so we add 500: $y = 500 + \sin(30t)$. Finally, this graph fluctuates above and below the average by 1 elk. But we want it to fluctuate by 50 elk, so we multiply the fluctuating term by 50. We have arrived at our model for the elk population (see Figure 20):

$$y = 500 + 50\sin(30t).$$

We can use this equation to predict the elk population at any time. For instance, in May (two months after March), the population will be $y = 500 + 50\sin(30 \cdot 2) = 543.3$ elk.

Passing over the question of what 0.3 elk might mean, there are still a couple of problems with this. We have no data to suggest that the elk population rises and falls according to the way the

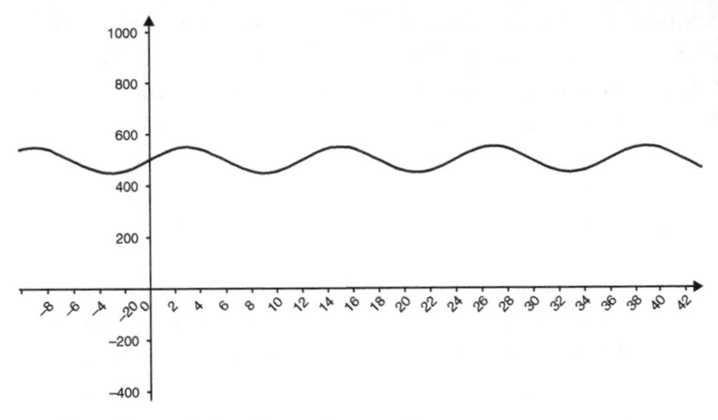

20. The elk population fluctuating over time.

sine function rises and falls. Clearly, elk considering whether or not to reproduce are not consulting the ratios of sides in a triangle on this romantic evening. But with extra evidence and more sophisticated ways of using the sine function as a building block for more complicated oscillations (some of which we'll see in Chapter 5), this approach can be extremely powerful.

This style of mathematics, so familiar to students, is an example of *analytic geometry*. It is usually said to have begun in the early 17th century with philosopher René Descartes of 'I think, therefore I am' fame, and his invention of *Cartesian coordinates*, our x and y axes. Like many stories in the history of mathematics, this is partly true in spirit but partly false in the specifics. Descartes did not clearly establish coordinate systems as we think of them today, and others were working in a similar vein at the same time. For instance, French mathematician Gilles de Roberval (1602–75) was thinking in this way about the *cycloid*, a favourite curve of 17th-century mathematicians. Imagine a fly lands on wheel AB at point A (Figure 21), and the wheel starts rolling to the right. If we set AC equal to half the circumference of the circle, the fly will arrive at its highest point D on the right just as point B

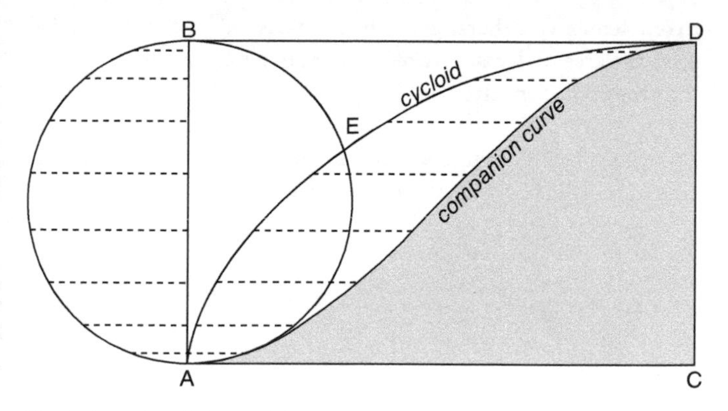

21. Roberval and the cycloid.

on the wheel reaches the ground at point C. The path travelled by the fly, the solid curve AED, is the cycloid.

Roberval was interested in the area under the cycloid. Calculus students might recognize this problem as an integration, but Roberval lived during an era when calculus was still very gradually taking shape. Roberval's argument isn't precisely calculus, but it has the flavour of its predecessors. If r is the circle's radius, then the area of rectangle $ABDC$ is $AB \cdot AC = 2r \cdot \left(\frac{1}{2} \cdot 2\pi r\right)$, since the right half of the circle AEB rolls along AC. So the rectangle's area is $2\pi r^2$, or four times the dashed semicircle. Now, notice that that semicircle is composed of horizontal lines. Take a copy of each horizontal line and slide it rightward so that its left edge touches the cycloid. The right edges of these lines form what Roberval called the 'companion curve'.

The companion curve cuts the rectangle into two equal parts, so the shaded area (one of those parts) is twice the area of the semicircle. But the two dashed regions each have exactly the same horizontal cross sections, so *they must have exactly the same area.* So the area between the cycloid and the companion curve is equal to one semicircle. Therefore, adding the dashed area to the shaded area, the area under the cycloid is equal to precisely three times the area of the semicircle.

The punchline? It turns out that the companion curve is part of a sine wave—the first time it was ever drawn.

Chapter 3
Building a sine table with your bare hands

Technological advances, so pervasive in almost every aspect of our modern lives, become mundane to us almost overnight. At the touch of a thumb, a smartphone retrieves a signal from any corner of the world through seemingly empty space. It processes that information with a built-in computer thousands of times more powerful than any of the original machines that used to fill up entire rooms. With this miracle, we can check whether anyone has scored a goal in the past minute in a football match taking place right now on the other side of the planet. Years ago this technology was a wonder; now it is ordinary, and we don't give it a second thought.

One of the oldest of these wonders, buried within the obscure functions of the calculator app on your smartphone, is the set of buttons that we used in Chapter 2 to find values of sines, cosines, and tangents. When we found the distance across the soccer field or the height of the tower, we may not even have noticed the mystery hidden there. How does a calculator find out, apparently effortlessly, that $\sin 33° = 0.5446$? There are no right triangles drawn inside of the calculator, so where did that number come from?

This story, spanning over two millennia, provides a motive for us to revisit some of the most common topics that we encounter in school trigonometry. Like any good story, there is a major twist about halfway through, which will be our focus in Chapter 5. Recall our 2nd-century BC Greek astronomer Hipparchus of Rhodes. His discovery of a method to find the eccentricity of the Sun's orbit around the Earth relied on being able to convert *arcs* or *angles* to *lengths* within a circle. He did this by constructing a table of chord lengths in a circle, now lost to history (although some have attempted to reconstruct it). We'll follow the steps Hipparchus might have taken to build his table, although we will use modern sines rather than Hipparchus' chords.

To begin, a few preliminaries. In Figure 22, imagine that we are watching the Sun rise starting in the east, passing directly overhead, and sinking in the west. (Usually the Sun rises at an angle, but let's assume for the moment that we're at the Earth's equator at an equinox.) The circle has a radius of one astronomical unit. Then the Sun's height above the ground is equal to the sine of its altitude θ. At sunrise, θ is equal to 0 and the Sun is not above the ground at all; therefore, $\sin 0° = 0$. As the Sun gets

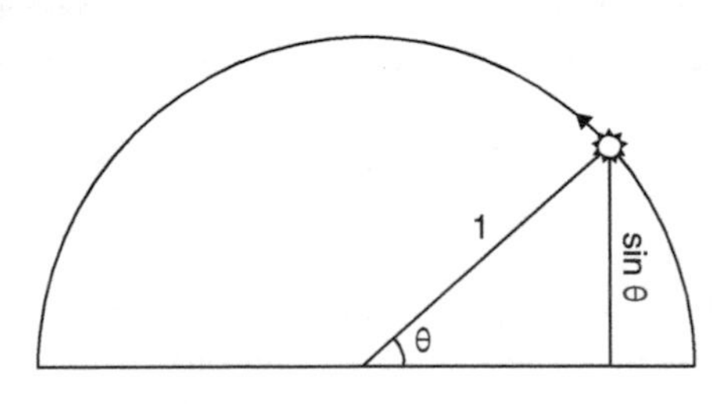

θ	$\sin\theta$
0°	0
1°	
⋮	⋮
89°	
90°	1

22. The beginning of our sine table.

higher in the sky, its altitude increases gradually. At noon, $\theta = 90°$, and the Sun directly above us is one astronomical unit above the ground. Therefore $\sin 90° = 1$. We insert these entries into our table, which leaves eighty-nine entries yet to be found: $\sin 1°, \sin 2°, \ldots, \sin 89°$.

Some of these sines are easier to determine than others. We begin as the historical astronomers did, by embedding various regular polygons in the unit circle. In Figure 23 we use a hexagon, which we divide into six triangles. The angles at the centre of the circle are each 60°, and since the triangles are all equilateral, all of their sides are 1 unit in length. If we cut the rightmost equilateral triangle in half, we get the bold right triangle in the diagram. The angle at its left corner is 30°; the hypotenuse is 1 unit; and the opposite side is exactly ½. So $\sin 30° = \frac{1}{2}$, and we have a new entry to place in our sine table.

If we embed a square rather than a hexagon into our circle (Figure 24), we can find $\sin 45°$. In the bolded triangle we know that the hypotenuse has length 1. The other two sides have

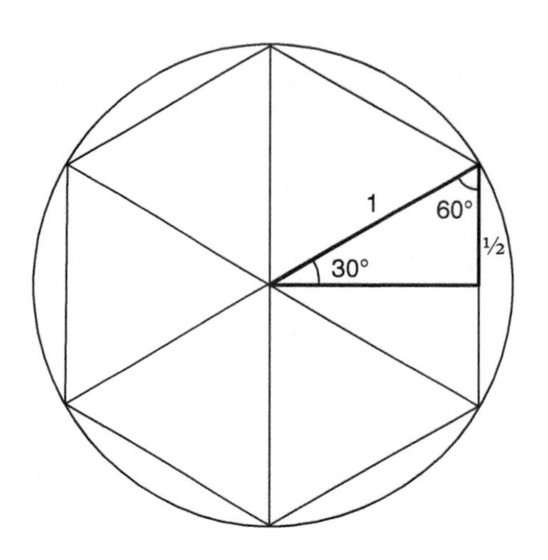

23. $\sin 30° = 0.5$.

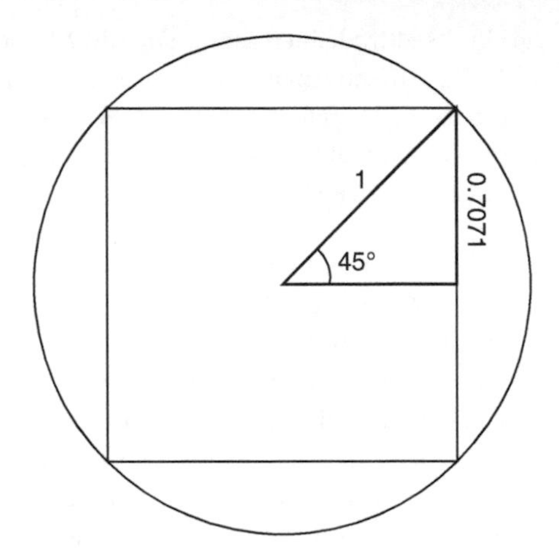

24. $\mathbf{\sin 45° = \sqrt{1/2}}$**.**

lengths equal to $\sin 45°$, which we call x. Then by Pythagoras we have $x^2 + x^2 = 1$, which gives us $x = \sqrt{\frac{1}{2}} = 0.7071$. Another entry for our table!

Finally, a bit of Pythagorean magic gives us one more easy sine value. Returning to Figure 23, let's consider the angle at the top right of the triangle; its value is 60°. We already know the hypotenuse is 1 and the vertical side is $\sin 30° = 0.5$. Since the base of the triangle is the side opposite to the 60° angle, we have

$$\sin 60° = \sqrt{1^2 - 0.5^2} = \sqrt{3}/2 = 0.8660.$$

These are the sine values commonly encountered in school. It's time to go beyond.

Finding more efficient ways to work: ancient astronomy meets modern computer graphics

We now have the sines of 0°, 30°, 45°, 60°, and 90°. If we're going to complete our table by finding the other eighty-six sines one entry at a time, this book will no longer be a *Very Short Introduction*. Thankfully, there's a better way. Imagine that we had a formula that would allow us to take the sines of any two given angles as inputs, and return to us the sine of the *sum* of those two angles. We could use it immediately to find $\sin 75°$, since $30° + 45° = 75°$; but probably we will be able to use it to find all sorts of other sine values as well. This seems likely to be much more productive than the one-at-a-time approach.

In the bottom left corner of Figure 25, angles α and β have been put together to form the larger angle $\alpha + \beta$. We set the length of OA to 1, so that AE is equal to the quantity we want to find, $\sin(\alpha + \beta)$. As one can see from the extra lines drawn in the diagram, AE can be broken into two parts, AD and DE. Let's find the lengths of those two pieces separately. First, notice that the hypotenuse of right triangle OAB is equal to 1, so $AB = \sin\beta$ and $OB = \cos\beta$. Now, consider right triangle OBC. In this triangle, $\sin\alpha = \frac{BC}{OB} = \frac{BC}{\cos\beta}$, which gives us $BC = \sin\alpha\cos\beta$.

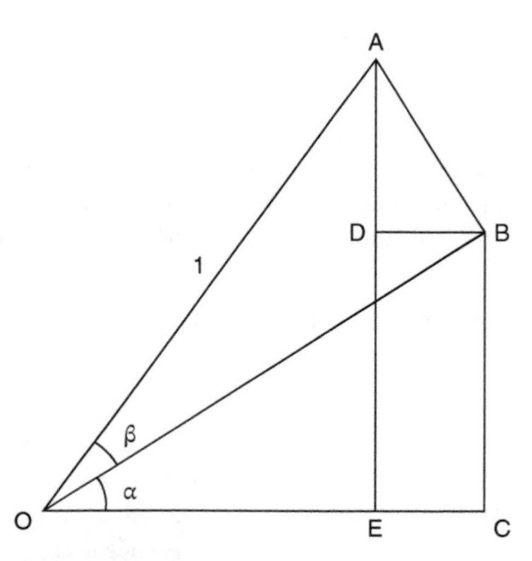

25. The sine angle sum formula.

But $BC = DE$, so now we know the length of one of the two parts of AE. Next, consider right triangle DAB. In this triangle, $\measuredangle DAB = \alpha$ (we'll leave this as a puzzle to discover why); hence $\cos\alpha = \frac{AD}{AB} = \frac{AD}{\sin\beta}$, which gives us the second part $AD = \cos\alpha \sin\beta$. Put these two pieces together, and we have our conclusion, called an 'identity':

$$\text{SINE ANGLE SUM FORMULA}: \\ \sin(\alpha+\beta) = \sin\alpha\cos\beta + \cos\alpha\sin\beta$$

You might consider how to use or modify Figure 25 to generate three other, closely related identities:

$$\text{COSINE ANGLE SUM FORMULA}: \\ \cos(\alpha+\beta) = \cos\alpha\cos\beta - \sin\alpha\sin\beta$$

$$\text{SINE ANGLE DIFFERENCE FORMULA}: \\ \sin(\alpha-\beta) = \sin\alpha\cos\beta - \cos\alpha\sin\beta$$

$$\text{COSINE ANGLE DIFFERENCE FORMULA}: \\ \cos(\alpha-\beta) = \cos\alpha\cos\beta + \sin\alpha\sin\beta$$

The symmetries in these formulas are uncanny. In somewhat different forms (recall that ancient Greek astronomers used chords rather than sines and cosines), they go back at least as far as Claudius Ptolemy's *Almagest* (AD 140), where he used them just as we are using them here—to build a trigonometric table.

A slightly more modern approach to the sine and cosine angle sum formulas

For readers who know a bit of linear algebra, we can encounter these formulas another way using the mathematics of computer graphics. Let's take the near-photorealistic image of my cat in Figure 26, a collection of vertices connected by straight lines, and x and y axes placed so that they intersect at the centre of the screen. As part of a poorly conceived PowerPoint presentation, I want to spin my cat on the screen to attract the viewer's attention. I will need to rotate each of my vertices through every possible angle θ, calculate the coordinates of all the rotated points, and connect the new points with straight lines.

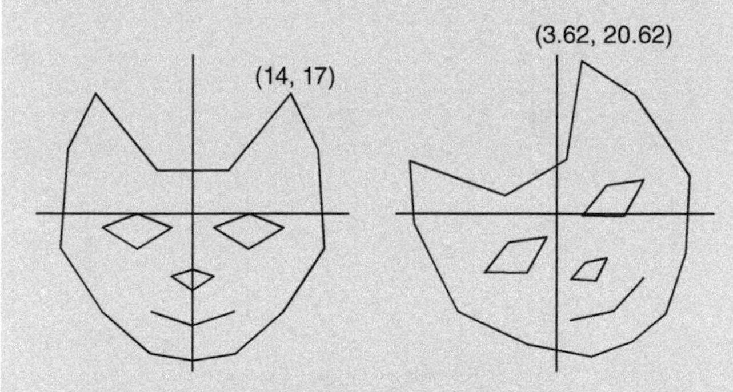

26. Rotating my cat 30° anticlockwise.

To work out the matrix transformation that performs the rotation, we need to know what the transformation does to the vectors $\mathbf{i}=\begin{bmatrix}1\\0\end{bmatrix}$ and $\mathbf{j}=\begin{bmatrix}0\\1\end{bmatrix}$. We can see this effect in the unit circle diagram in Figure 27: our matrix multiplied by **i** needs to be $\begin{bmatrix}\cos\theta\\ \sin\theta\end{bmatrix}$, and our matrix multiplied by **j** needs to be $\begin{bmatrix}-\sin\theta\\ \cos\theta\end{bmatrix}$. The matrix that accomplishes this task is

$$\begin{bmatrix}\cos\theta & -\sin\theta\\ \sin\theta & \cos\theta\end{bmatrix}.$$

To rotate the cat 30° anticlockwise, we multiply the coordinates of each of the vertices by this matrix. For instance, the tip of the cat's ear at (14, 17) moves to

$$\begin{bmatrix}\cos 30^\circ & -\sin 30^\circ\\ \sin 30^\circ & \cos 30^\circ\end{bmatrix}\begin{bmatrix}14\\17\end{bmatrix}=\begin{bmatrix}3.62\\21.72\end{bmatrix}.$$

(continued)

Continued

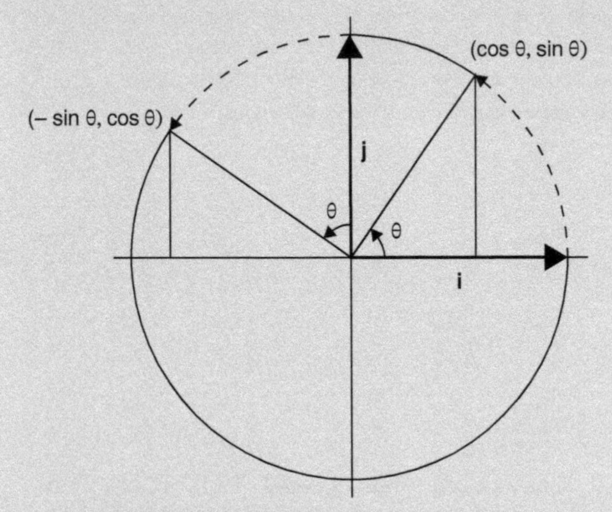

27. Rotating the basis vectors i and j.

What does this have to do with the sine angle sum and difference laws? Let's consider the matrix of a rotation through the sum of angles α and β. We know it will be

$$\begin{bmatrix} \cos(\alpha+\beta) & -\sin(\alpha+\beta) \\ \sin(\alpha+\beta) & \cos(\alpha+\beta) \end{bmatrix}.$$

But we also can think of it as a rotation first through angle α, and then through angle β. So the matrix will also be

$$\begin{bmatrix} \cos\beta & -\sin\beta \\ \sin\beta & \cos\beta \end{bmatrix} \cdot \begin{bmatrix} \cos\alpha & -\sin\alpha \\ \sin\alpha & \cos\alpha \end{bmatrix} =$$
$$\begin{bmatrix} \cos\alpha\cos\beta - \sin\alpha\sin\beta & -\sin\alpha\cos\beta - \cos\alpha\sin\beta \\ \sin\alpha\cos\beta + \cos\alpha\sin\beta & \cos\alpha\cos\beta - \sin\alpha\sin\beta \end{bmatrix}.$$

As if by magic, we find the cosine angle sum formula in the top left entry, and the sine angle sum formula in the bottom left.

It's time to use our hard-won formulas to generate some more sine values; after all, that is why we worked to get them. Since $30^\circ + 45^\circ = 75^\circ$,

$$\begin{aligned}\sin 75^\circ &= \sin 30^\circ \cos 45^\circ + \cos 30^\circ \sin 45^\circ \\ &= \sin 30^\circ \sin 45^\circ + \sin 60^\circ \sin 45^\circ \\ &= 0.5 \times 0.7071 + 0.8660 \times 0.7071 \\ &= 0.9659.\end{aligned}$$

We can use the sine angle difference law to find $\sin 15^\circ$ as follows:

$$\sin 15^\circ = \sin(45^\circ - 30^\circ) = \sin 45^\circ \cos 30^\circ - \cos 45^\circ \sin 30^\circ = 0.2588.$$

We now have a small but satisfying table of the sines of all the multiples of 15° (Figure 28). This small table was an established part of trigonometry for almost a millennium. It appeared first in the work of astronomer Brahmagupta, from 7th-century AD

θ	$\sin \theta$
0	0
15°	0.2588
30°	0.5
45°	0.7071
60°	0.8660
75°	0.9659
90°	1

Arcus.	Sinus.
90	600000000
30	300000000
60	519615242
45	424264069
15	155291427
75	579555496

28. Our table of the sines of the *kardajas* at left, and Regiomontanus' table at right, from his *Compositio tabularum sinuum rectorum*. His table is a little more accurate than ours. Regiomontanus' table is given in the order that he computed the values, not in ascending order of the arcs. Instead of a unit circle, he used a circle of radius *R*=600,000,000; this allowed him to be accurate without needing decimal fractions, which were not in use at his time.

Bhinmal, India. His table didn't look much like ours; rather than our unit circle he used a circle with radius 150. Also, his table wasn't displayed in rows and columns, but recited in more easily memorized verse; much of Indian astronomy took place within an oral tradition. The 15° table travelled far. It found its way through the Middle East into Muslim Spain, and from there into medieval Europe. The 15° increments came to be known as the *kardajas*, from the Persian word for 'section'. One of the final appearances of a *kardaja* table was in the work of early Renaissance Viennese astronomer Regiomontanus, reproduced in Figure 28.

Coming up golden

Unfortunately, we are now stuck. Since all the angles in our table are multiples of 15°, adding or subtracting them from each other won't generate any new sine values. We need a new idea.

That idea will prove to be golden. Recall that we started finding sine values by considering the square and the regular hexagon. But we skipped over the pentagon, so we turn to it now.
In Figure 29, our pentagon has side lengths equal to 1. If we draw lines from the top vertex A to the vertices C and D at the bottom, we have formed the shaded *golden triangle*. It may not look particularly lustrous at the moment, but sometimes it takes a little digging to find a nugget. If we draw CF at a 36° angle to CD, we've broken our golden triangle into a larger isosceles triangle AFC, and a smaller copy of the golden triangle, CDF.

Let's find the lengths of the sides of our golden triangles. The large shaded triangle's base has length 1; we call the other two side lengths $\varphi = AC = AD$. In the smaller triangle, the two longer sides CD and CF are 1. We can find the shorter side DF cleverly as follows. Triangle AFC is isosceles, so AF must be equal to CF, which is 1. But AD is equal to φ, so $DF = \varphi - 1$. Since the two

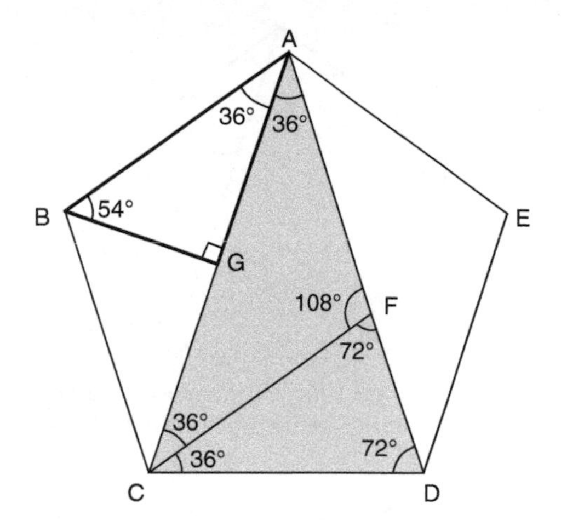

29. Using the regular pentagon and the golden ratio to find $\sin 36°$ and $\sin 54°$.

golden triangles are similar, the ratios of their sides must be equal. Hence,

$$\frac{\varphi}{1} = \frac{1}{\varphi - 1}.$$

Cross-multiplying, we get the quadratic equation $\varphi^2 - \varphi = 1$, whose solution is one of the most remarkable numbers in all of mathematics, the *golden ratio*:

$$\varphi = \frac{-1 + \sqrt{5}}{2} = 1.61803\ldots$$

What's so remarkable about this number? For one, it appears in a dazzling array of areas of mathematics, natural phenomena, works of art, and architecture. We can see it already in our pentagon. If we fill it with large golden triangles to form a pentagram (Figure 30), many of the pairs of line segments within the pentagram form the golden ratio.

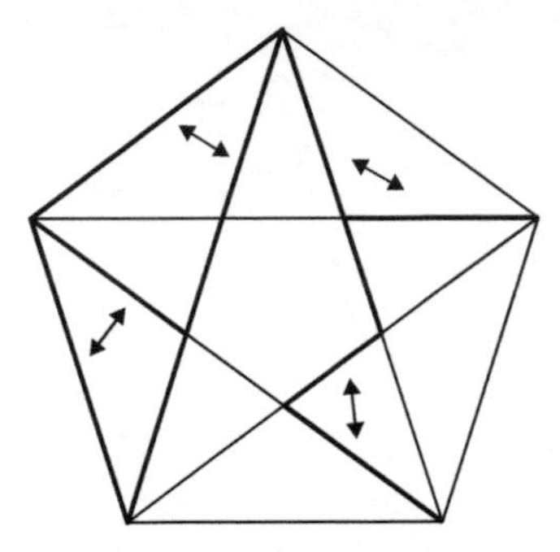

30. Several instances of the golden ratio in the pentagram. Each arrow refers to the entire lengths of the bold-faced segments to which they point.

φ isn't just in geometry; it's also in arithmetic. The *Fibonacci sequence* is generated by adding the two previous numbers in the sequence to get the next one:

$$1, 1, 2, 3, 5, 8, 13, 21, 34, 55, 89, \ldots$$

The ratio of each number in the sequence to the previous number gets closer and closer to φ as you go further and further along.

We find φ in the natural world as well. Count the number of scales in a row on the outside of many pine cones (Figure 31): in one direction you get a number in the Fibonacci sequence; in the other direction you get the next number in the sequence. Natural growth patterns often seem to employ the golden ratio, although specific claims can be controversial: some people see more than others. Equally fiercely debated are the aesthetic benefits of using the golden ratio in paintings, sculptures, building designs, and even musical compositions.

Again we seem to have drifted away from the topic of constructing sines; but again, surprisingly, we are on its doorstep. Consider the

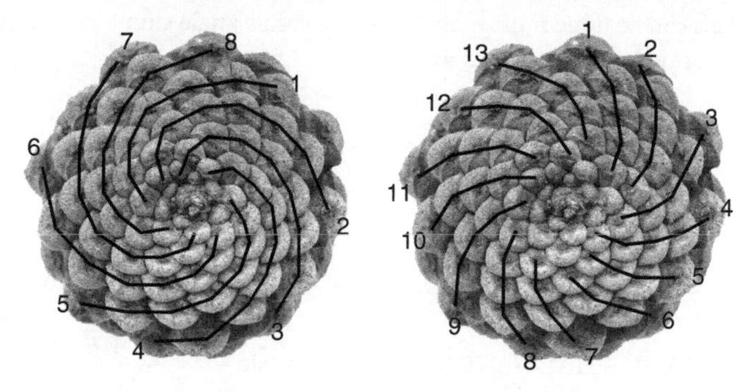

31. Counting the scales in a pine cone in two different directions.

bolded right-angled triangle on the left of Figure 29, with an angle of 54° at B. Since its hypotenuse is 1, we have

$$\sin 54° = AG = \frac{AC}{2} = \frac{\varphi}{2} = 0.8090;$$

and using the Pythagorean theorem on this same triangle,

$$\sin 36° = \sqrt{1-(\varphi/2)^2} = 0.5878.$$

Thus the golden ratio is embedded within every sine table. Even better, these two new sines unlock a cascade of other sine values. From 30° and 36°, the sine angle difference formula gives us

$$\sin 6° = \sin(36° - 30°) = \sin 36° \cos 30° - \cos 36° \sin 30° = 0.1045.$$

From sin 6°, we can repeatedly apply the sine angle sum formula to calculate the sines of all the multiples of 6°. From here we could use 42° and 45° with the difference formula to get $\sin 3° = 0.05234$. But another approach will garner us a bonus result along the way. Obviously 3° is half of 6°, so let's find a sine half-angle formula.

This can be done most easily using the cosine angle sum formula, substituting $\frac{\theta}{2}$ for both α and β:

$$\cos\theta = \cos\left(\frac{\theta}{2}+\frac{\theta}{2}\right) = \cos\frac{\theta}{2}\cos\frac{\theta}{2} - \sin\frac{\theta}{2}\sin\frac{\theta}{2} = \cos^2\frac{\theta}{2} - \sin^2\frac{\theta}{2}.$$

But $\cos^2\frac{\theta}{2} = 1 - \sin^2\frac{\theta}{2}$, so the entire right side is equal to $1 - 2\sin^2\frac{\theta}{2}$. Solving this equation for $\sin\frac{\theta}{2}$, we arrive at our goal:

$$\text{SINE HALF-ANGLE FORMULA: } \sin\frac{\theta}{2} = \sqrt{\frac{1-\cos\theta}{2}}$$

With this formula we can seal off a loose end from Chapter 2. Rearranging this equation, we find that $1 - \cos\theta = 2\sin^2\theta$; but $1 - \cos\theta$ is the versed sine, thereby confirming that the versed sine is always positive.

Now, from $\sin 6°$ we can apply this formula as many times as we like, to get $\sin 3° = 0.05234$, and $\sin\frac{3°}{2} = 0.02618$, and $\sin\frac{3°}{4} = 0.01309$, and so on. Or we can apply it to any other angle for which we have the sine. For instance, from 36° we find $\sin 18° = 0.3090$. However, let's end this section with a flourish by finding $\sin 18°$ in a surprising way.

Many have heard the name Euclid of Alexandria, although perhaps not so many know who he was. Working in the 3rd century BC, he was the author of the greatest mathematics textbook of all time, the *Elements*. Here Euclid compiled the knowledge of his predecessors into a single formal logical structure, establishing even today, two millennia later, the standard for how mathematics should be conceived and presented. In the last of the *Elements*' thirteen 'books' (i.e. chapters), Euclid described geometry in three dimensions (Figure 32). This must have been a very new subject, since only a century earlier Plato had despaired to report how little was known.

32. From the first English translation of Euclid's *Elements* by Henry Billingsley, with pop-up diagrams in the books on three-dimensional geometry.

In preparation for working with three-dimensional solids like tetrahedra, cubes, and dodecahedra, Euclid proved a couple of startling facts about regular polygons inscribed in a circle. The first is that the ratio between the side of the hexagon and the side of the decagon is precisely equal to φ, the golden ratio. The second states that if we put together the sides of the pentagon, hexagon, and decagon, they form a perfect right triangle (Figure 33). In the unit circle, the side of a pentagon is $2\sin 36°$, the side of the hexagon is $2 \sin 30°$, and the side of the decagon is $2\sin 18°$. Therefore

$$(2\sin 36°)^2 = (2\sin 30°)^2 + (2\sin 18°)^2,$$

from which we find $\sin 18° = 0.3090$.

Shifting our ground, from geometry to algebra

Very well; we now have the sines of every multiple of 3°. But just as when we had arrived at the *kardaja* multiples of 15°, we are

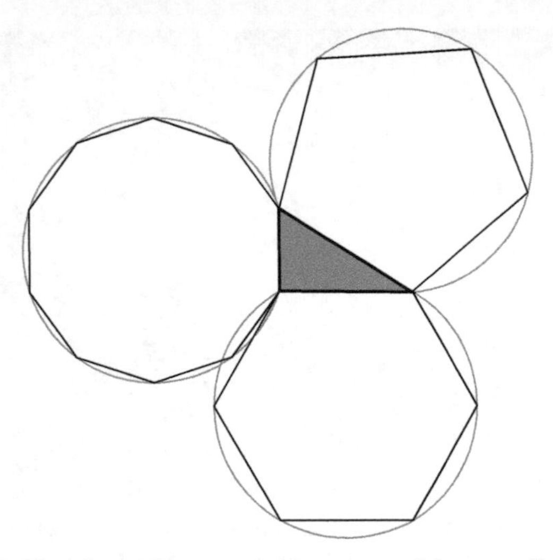

33. The sides of a regular pentagon, hexagon, and decagon, all inscribed in circles of the same size, form a right triangle.

again at a loss to fill in any other entries. We should not lose hope right away. Since trigonometric identities were able to take us past the *kardaja* barrier before, why not this time?

Unfortunately, one cannot always have what one wants. Ancient Greek geometers were able to perform remarkable feats with only the basic tools at their disposal: the straightedge and compass. However, three of the simplest tasks, using only these tools, remained forever out of their grasp. The first is to *construct a square with the same area as a given circle* (Figure 34). This problem was finally proved to be impossible only in 1882 by Ferdinand von Lindemann; the phrase 'squaring the circle' has entered into some parts of popular culture meaning to attempt a hopeless undertaking. To *construct a cube with precisely double the volume of a given cube* was proved impossible by Pierre Wantzel in 1837. (Simply doubling the sides of the original cube will not do. Your new cube would be eight times—not twice—the volume of the given cube.)

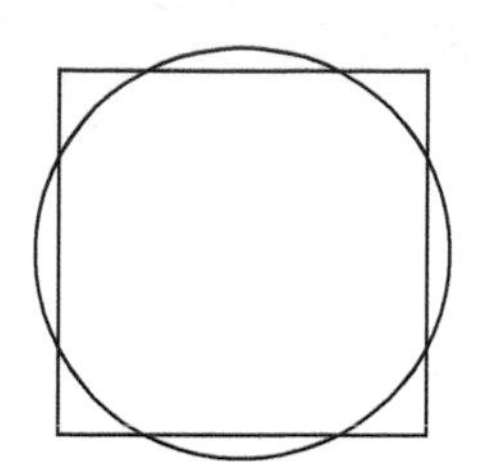
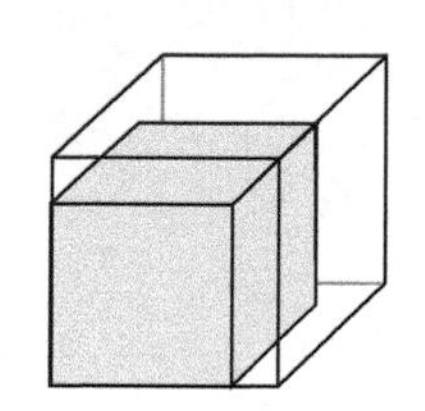
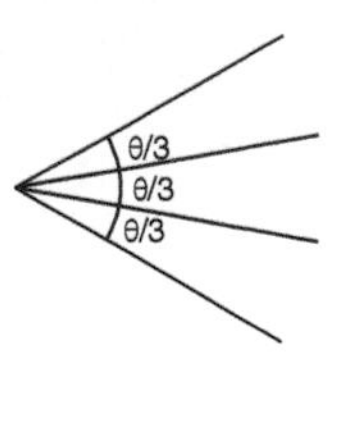

34. The three classical Greek problems: squaring the circle, doubling the cube, and trisecting the angle.

The third problem is also deceptively simple: *given an angle, divide it into three equal parts*. Euclid had shown how to bisect an angle already in the ninth of the 465 propositions in his *Elements*, but he was silent about trisecting an angle. Wantzel finally proved it to be impossible in 1837, in the same article where he dealt with doubling the cube. In all three cases, it was the emergence of new concepts in what is now called abstract algebra that resolved the mystery. Before then, it was generally assumed that the problems could not be solved, but that did not—and does not, even today—prevent people from making the attempt.

We already constructed an angle of 3° and calculated its sine, which we give here to a few more decimal places: 0.0523359562. To work from this value down to sin 1° using geometry requires us to trisect our 3° angle. Now, we've already been able to trisect *some* angles; for instance, we know both sin 45° and sin 15°. But we can't trisect *every* angle, and unfortunately 3° is one of those angles that is impossible to trisect.

The geometrically minded ancient and medieval astronomers had no choice: if they wanted a sine table, they had to change how they thought about the problem. Many found some way to capture the inaccessible sine between upper and lower bounds that were very close to each other. The earliest record we have of such an effort is in Claudius Ptolemy's *Almagest*—the same book in which equivalents to the sine angle sum and

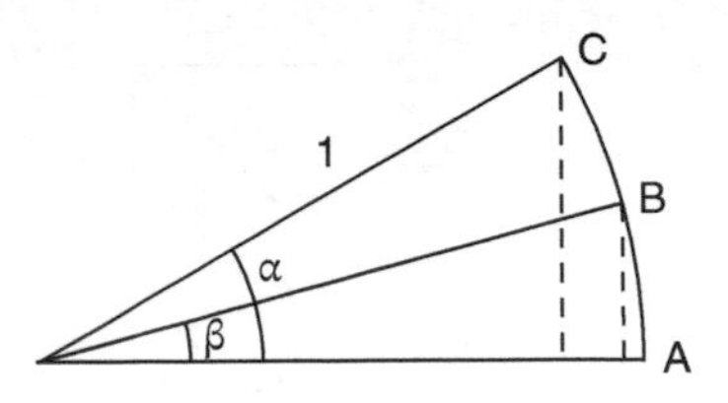

35. Ratios between sines of small angles.

difference laws first appeared. Ptolemy's method, as follows, works remarkably well.

A few of pages back, you may have noticed that our value for $\sin\frac{3^\circ}{2} = 0.02618$ appears to be exactly double that of $\sin\frac{3^\circ}{4} = 0.01309$. That isn't quite true, as we can see from the more precise values 0.0261769 and 0.0130896. However, the near coincidence can be explained and exploited using Figure 35. Suppose we have two small angles α and β in a unit circle; their sines are the two dashed lines. If (for instance) α is twice β, then $\sin\alpha$ will be larger than $\sin\beta$. But it won't quite be twice as large, because the increase in the sine as we move from B to C is a little less than the increase as we move from A to B. In other words, if $\alpha > \beta$, then

$$\frac{\alpha}{\beta} > \frac{\sin\alpha}{\sin\beta}.$$

Now we turn from the explanation to the exploitation. Plug $\alpha = \frac{3^\circ}{2}$ and $\beta = 1^\circ$; then plug in $\alpha = 1^\circ$ and $\beta = \frac{3^\circ}{4}$. Solving for $\sin 1^\circ$ in both cases, we get

$$0.0174513 = \frac{2}{3}\sin\frac{3^\circ}{2} < \sin 1^\circ < \frac{4}{3}\sin\frac{3^\circ}{4} = 0.0174528.$$

These bounds are very close to each other, so we can say with confidence that $\sin 1^\circ \approx 0.017452$. Using this value, we can go ahead and calculate the rest of the entries in our sine table, and rest assured that the entries should be reliable to five decimal places, more or less.

But what if we want more?

We might try to revisit our approach to the sine half-angle formula, where we wrote $\cos\theta$ as $\cos\left(\frac{\theta}{2}+\frac{\theta}{2}\right)$ and expanded using the cosine angle sum law. If we're lucky, we might be able to derive a sine one-third angle formula in a similar way. Here's what happens: applying the sine angle sum law to $\sin\theta=\sin\left(\frac{\theta}{3}+\frac{\theta}{3}+\frac{\theta}{3}\right)$, after some algebra we arrive at

$$\sin\theta=3\sin\frac{\theta}{3}-4\sin^3\frac{\theta}{3},$$

which gives $\sin 3°=3\sin 1°-4\sin^3 1°$. Since we already know $\sin 3°$, this gets us tantalizingly close to $\sin 1°$. But now we're in trouble: this is a cubic equation. We may have solved *quadratic* equations in school, but you would have been a rare student indeed if you also solved *cubics*.

The first time someone tried to use our formula to solve for $\sin 1°$, the cubic equation was still well over a century away from being solved by anyone. The setting was early 15th-century Samarqand, where Timurid sultan Ulugh Beg seemed to be just as interested in running his astronomical observatory and research institute as in running his empire. His star scientist was Jamshīd al-Kāshī, one of the most brilliant computational minds of all time. Al-Kāshī came up with a clever method to find $\sin 1°$ to as much precision as one wants using our cubic equation, not long before his untimely death in 1429. Several of his colleagues and successors expanded on his work and found variations on his method; the one we shall see here is attributed to Ulugh Beg himself.

Let x represent the sought value of $\sin 1°$. Our cubic equation becomes

$$0.0523359562 = 3x - 4x^3.$$

Although we can't solve this equation for x, we can at least isolate one of the occurrences of x by rearranging it:

$$x = \frac{0.0523359562 + 4x^3}{3}.$$

We could make a naïve guess that x is somewhere around $\frac{1}{3} \cdot \sin 3° = 0.0174453187$, which is close but not nearly close enough. Let's examine a picture of this situation. Consider *both* sides of the equation as functions of x, in other words, $y = x$ and $y = \frac{0.0523359562 + 4x^3}{3}$. These two functions are graphed (not to scale) in Figure 36; they cross at the value of x that we're seeking. We see in the figure that our guess of 0.0174453187 was too small, which we can confirm numerically: $\frac{0.0523359562 + 4(0.0174453187)^3}{3} = 0.0174523978$ is a bit bigger than 0.0174453187.

Inspired by al-Kāshī, Ulugh Beg made a clever move. The value 0.0174523978 that we just obtained is the y value (or the height of the curve) corresponding to our original x value in the figure. Imagine reflecting this y value across the line $y = x$, thereby turning it into an x value (illustrated with bold arrows). Since the curve in the figure has a gentle slope, our new x value will be quite a bit closer to the desired intersection point.

No need to stop there. Repeating the process, we insert our new estimate back into the equation of the curve: $\frac{0.0523359562 + 4(0.0174523978)^3}{3} = 0.0174524064$.
Again, we have moved a little closer. Repeat again: $\frac{0.0523359562 + 4(0.0174524064)^3}{3} = 0.0174524064$. To the ten

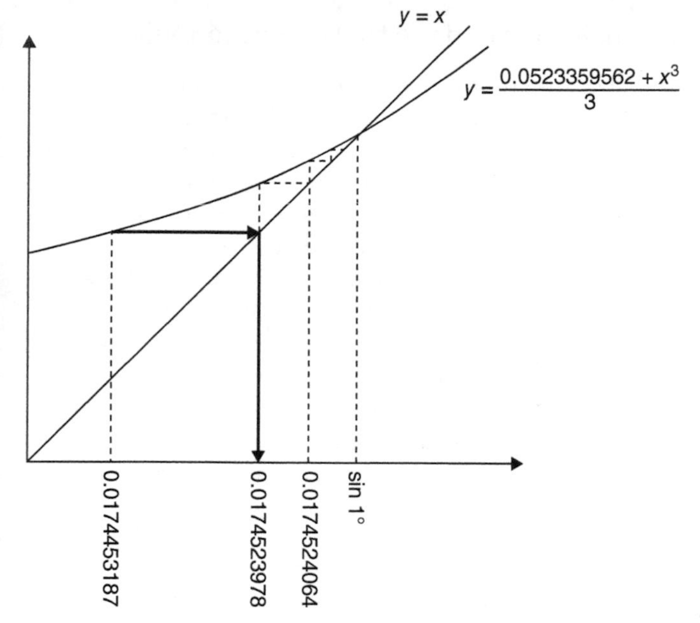

36. Ulugh Beg's iteration process to find sin 1°.

decimal places we've been using, this new guess is identical to the previous one. In fact, this quantity is equal to sin 1° to all ten places.

This process is an example of *fixed-point iteration*, one of the earliest methods used by numerical analysts to approximate solutions to equations that can't be solved directly. The method had been used already in ancient Greece to find square roots and in India to solve complicated astronomical problems. Now, neither the Greeks, Indians, or Ulugh Beg would have visualized an intersection of two curves as we have done here. Even the notion of a graph is a much later invention, from the 17th century. This fact makes their accomplishments, likely achieved using only numerical intuition, all the more marvellous.

Extending beyond the triple-angle formula

Now that we have an extremely good estimate for $\sin 1°$, we can use the sine sum and difference laws to fill in the rest of our sine table. But as needs for precision increased, a table of sines with arguments separated by 1° soon became insufficient: the errors caused by interpolating between these entries was just too great. In 16th-century Europe, sine tables started to be constructed with increments of 10′ (ten minutes, or sixtieths of a degree) or even $1'$. Georg Rheticus and Valentin Otho's 1596 *Opus palatinum* took this to an extreme, with entries for every 10″ (ten 3,600ths of a degree) of arc! The magnitude of Rheticus' and Otho's project is obvious from a quick glance at his book: the tables fill almost 750 very large pages, and to complete them required the assistance of five human computers working full time for twelve years.

Suppose, then, that we want to find sin 1′ from $\sin 1°$, which would allow us to take our table down from $1°$ increments to 1′ increments. Our trisection method immediately gets us sin 20′; the half-angle formula gives sin 10′ and then sin $5'$. But now, as before, we are stuck—unless we can find a formula for $\sin 5\theta$. We might anticipate having to solve a fifth-power equation, but that would be a small price to pay for a greatly improved sine table.

Enter François Viète (1540–1603). A lawyer by trade and eventually privy councillor to two kings of France, Viète lived during a dangerous time of strife between the Catholics and Protestants. Although he was officially a Catholic, he sheltered and defended Protestants throughout his life, and paid a political price for his actions. His work in mathematics was a side interest; due to his wealth he was able to self-publish his findings. Today his name is associated with the development of symbolic algebra, the engine that underlies most of today's mathematics. However, his first published works were in astronomy—and, of course, trigonometry.

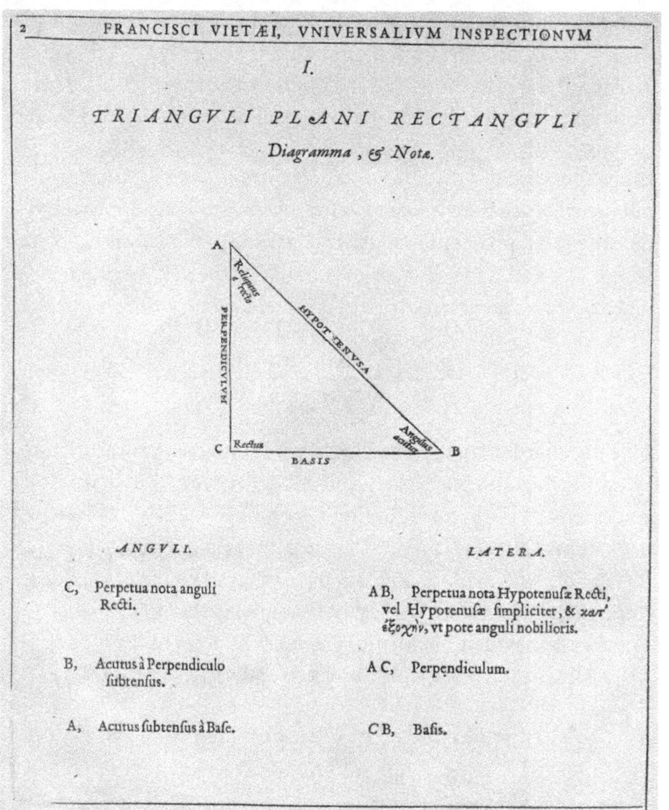
2 FRANCISCI VIETÆI, VNIVERSALIVM INSPECTIONVM

I.

TRIANGVLI PLANI RECTANGVLI

Diagramma, & Notæ.

ANGVLI.

C, Perpetua nota anguli Recti.

B, Acutus à Perpendiculo ſubtenſus.

A, Acutus ſubtenſus à Baſe.

LATERA.

A B, Perpetua nota Hypotenuſæ Recti, vel Hypotenuſæ ſimpliciter, & κατ' ἐξοχὴν, vt pote anguli nobilioris.

A C, Perpendiculum.

C B, Baſis.

37. The first page of the text accompanying Viète's *Canon mathematicus*, where he lays out the parts of a right triangle.

Viète's first publication, his 1579 *Canon mathematicus seu ad triangula*, is one of the most beautifully laid-out mathematical research monographs you will ever see, almost a coffee table book (Figure 37). It includes trigonometric tables (including a rather strange one that represents all its sine values as fractions) and uses unique shorthand notations that foreshadowed his later work in symbolic algebra. In this book Viète cleverly approximated $\sin 1' = 0.000290882056$, a value that is accurate to all decimal places but the last. However, his method wasn't much different from what his predecessors had done, all the way back to Ptolemy.

In his much later work, *Ad angularium sectionum analyticen*, published more than a decade after his death, Viète approaches the sine value problem essentially the same way that al-Kāshī and Ulugh Beg had done, but he goes much further. He begins by determining a *recurrence relation* for $\sin n\theta$:

$$\frac{\sin\theta}{\sin 2\theta} = \frac{\sin((n-1)\theta)}{\sin((n-2)\theta) \cdot \sin n\theta}.$$

This peculiar formula contains within it immense power: if you know the formula for $\sin 2\theta$ and $\sin 3\theta$ (which by now we do), you can use it with $n = 4$ to find a formula for $\sin 4\theta$. Once you have this result, you can use the equation again to find a formula for $\sin 5\theta$. And again…and again…as far as you like. He does something similar for cosines, deriving formulas up to $\cos 21\theta$. The forms in which his formulas appeared differed from ours; we're content here to display modern equivalents:

SINE QUINTUPLE ANGLE FORMULA:

$$\sin 5\theta = 5\sin\theta - 20\sin^3\theta + 16\sin^5\theta$$

As we guessed, this turns out to be a fifth-power formula. Nevertheless, it is still solvable using an approximate method similar to the cubic solution we saw earlier. Now Viète can find $\sin 1'$: starting with our value for $\sin 18°$, we apply our quintuple angle formula, giving $\sin 3°36'$. From our value for $\sin 60°$ and the triple-angle formula, we get $\sin 20°$; trisect again to get $\sin 6°40'$, and then bisect to get $\sin 3°20'$. Apply the sine difference law to $3°36'$and $3°20'$, and we have $\sin 16'$. Finally, bisect four times, and we arrive at $\sin 1'$. Viète died before he could pursue the actual calculations, but three decades later

Henry Briggs carried it out in his massive trigonometric tables, the *Trigonometria Britannica.*

$$\text{SINE 21} \times \text{ANGLE FORMULA:}$$
$$\begin{aligned}\sin 21\theta = 21\sin\theta - 1540\sin^3\theta + 33\,264\sin^5\theta - 329\,472\sin^7\theta \\ +1\,793\,792\sin^9\theta - 5\,870\,592\sin^{11}\theta + 12\,042\,240\sin^{13}\theta \\ -15\,597\,568\sin^{15}\theta + 12\,386\,304\sin^{17}\theta - 5\,505\,024\sin^{19}\theta \\ +1\,048\,576\sin^{21}\theta\end{aligned}$$

I admit, that one was just for fun.

Now that we've finished calculating sine tables, the question arises how to calculate a cosine or a tangent table. These are easy: a cosine table is simply a sine table read backwards, and we can make a tangent table by dividing sines by cosines.

An excursion to India

The methods we've seen in this chapter are not how sines are calculated today; we'll see that in Chapter 5. Nor were they the only game in town in the ancient world. We turn next to India, where as early as the 5th century AD astronomers were thinking very differently about the same topics. Āryabhaṭa, from India's far north-east, devised a method of building a sine table that can be programmed into a spreadsheet in seconds and that can approximate countless sine values almost instantly. Like most writings of the time in India, his text was compressed so that it could be memorized—so compressed that it isn't quite clear what Āryabhaṭa was thinking. Explanations have been attempted for centuries, including by the famous 15th-century astronomer Nīlakaṇṭha. But they go beyond our scope (see Further Reading for more information).

Āryabhaṭa didn't work with the same sines that we do. He didn't use our unit circle; instead, he used the Indian circle with a radius

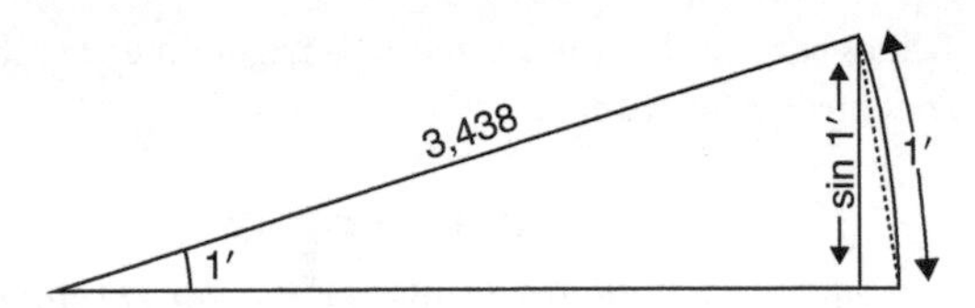

38. One minute of arc in Āryabhaṭa's trigonometric circle. The dashed line, one side in a 21,600-sided polygon, is the unit of distance. The vertical line, sin 1′, is just a shade less than one unit long.

of 3,438 units that we saw in Chapter 2 (see Figure 38). This seemingly strange choice was not made randomly. We divide the circle into 360°, and each degree consists of 60′, so there are 21,600′ in a circle. Now think of the circle as a regular polygon with 21,600 sides, each of length 1. Since the circumference is 21,600 units long, the radius is $21{,}600/2\pi \approx 3{,}438$ units. One advantage to choosing this radius is that the sine of a small arc is very close indeed to the arc itself. For instance, in a circle of radius $21{,}600/2\pi$, the sine of 1′ is 0.999999986.

Āryabhaṭa's description of how to compute sines is only one sentence long, which we summarize as follows: '*the difference between one sine and the next is equal to the previous difference, minus 1/225 of the previous sine*'. In other words, if we have n entries so far,

$$\Delta(n+1) = \Delta(n) - \frac{1}{225}\cdot(n\text{th sine}),$$

where $\Delta(n)$ is the difference between the nth sine and the sine just before it. This formula gives us the ability to compute any sine value from the previous value in the table. But we'll need the very first sine to start us off.

Thanks to Āryabhaṭa's choice of the radius of 3,438, we already have that first entry. His table gives sines for every 1/24th of a right angle, or every 3¾°. Since a 3,438 radius guarantees that sines of small angles are very close to the angles themselves, his first sine value is simply $\sin(3\tfrac{3}{4}°) = \sin(225') \approx 225$, which is accurate to the nearest whole number. Now we can use Āryabhaṭa's equation to start filling in the table (see Figure 39). The first difference $\Delta(1)$ is $225 - 0 = 225$. To find $\Delta(2)$ we subtract $\frac{1}{225} \cdot \sin 3¾°$ from 225, getting 224. Finally, add the 224 to our first sine value (225) to get the second entry in the table: $\sin 7½° = 449$. Repeat as many times as you like. If you have a spreadsheet program, give this method a try. It takes just a minute to set up, and it produces accurate sine values amazingly quickly.

Angle	*Sine values*	*Differences*
0°	0	
3¾°	225	225
7½°	**449**	**224**
⋮	⋮	⋮

39. The first couple of rows of Āryabhaṭa's sine table. If you have the sine and difference values in a given row (here for 3¾°), use Āryabhaṭa's equation to combine those values into a new difference (solid arrows). Then add the new difference to the previous sine value to generate the next sine (dotted arrows).

From India, to radians

The idea to choose an angular measurement system so that sines of small angles are approximately equal to the angles themselves is precisely the idea behind *radians*, the other way that angles are measured today. For radians, however, we leave the unit circle as it is. Instead we measure our arcs along the outside of the circle using the radius as the unit of length (see Figure 40). Since the circle's circumference is 2π, we then have 2π radians in a complete circle rather than 360 degrees.

(*continued*)

Continued

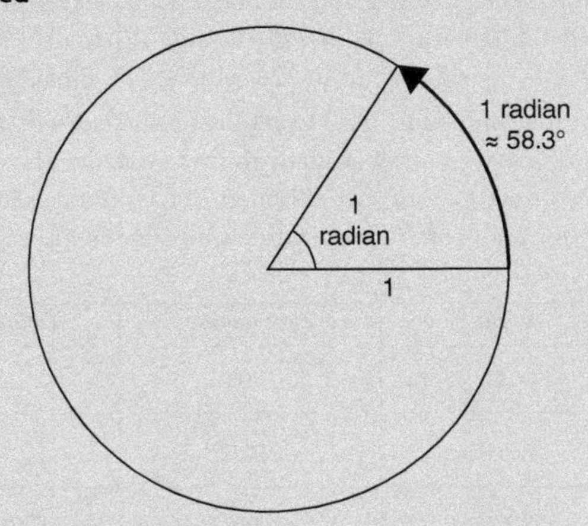

40. Radian measure.

Measuring angles and arcs in this way is convenient in calculus. If you've studied the subject, you may recall that the derivative of the sine is the cosine:

$$[\sin x]' = \cos x.$$

This is only true if you are using radian measure. In degrees, it's slightly messier:

$$[\sin x]' = \frac{\pi}{180} \cos x.$$

This doesn't seem like much of a difference. But if we use degree measure in complicated real-world situations, $\frac{\pi}{180}$ terms end up littered all over the place. One *could* do calculus using degree measure, but one would *prefer not to*.

By the way, the relation $[\sin x]' = \cos x$ is hidden, sort of, in Āryabhaṭa's table. If you have done the spreadsheet exercise, notice that the entries in the difference column gradually decrease faster and faster as you work your way down the table. If you plot them, you will find a nearly perfect cosine curve.

Chapter 4
Identities, and more identities

In Chapter 3 we saw a number of identities that helped us to construct a sine table without any mechanical aids. These are the ones that most of us learned in school—the sine sum and difference laws, the half- and double-angle formulas, and so on. But they are just the beginning. The world of trigonometry is full of identities: some of them extremely useful, others beautiful, and a few that are simply bizarre. In this chapter we shall take a tour of the menagerie of identities, viewing a little from each of these categories. You may have seen some of them before, but very few of us have seen all of the obscure, wonderful, and curious identities that inhabit the trigonometric zoo. Our first two examples are known as *triangle identities*, because they refer to angles and lengths in a given triangle.

The Law of Sines

Early 15th-century Venice might have been the very definition of the word 'bustling'. The centre of world trade, it was the stage for a dazzling mix of cultures hawking their wares, representing almost every society from what are now Europe and the Middle East. The Venetian merchant fleet transported goods over a region ranging from the Black Sea to Spain. The navigators of these vessels (Figure 41) laid their courses around the Mediterranean Sea using a variety of techniques preserved today in their notebooks. One of

41. From Michael of Rhodes' navigational notebook.

these sailors, Michael of Rhodes, included in his notebook a method known as the *marteloio*—effectively, a unique form of trigonometry. There was no established practice of trigonometry in Venice at the time; no one knows precisely where Michael got it. It might have come originally from Fibonacci a couple of centuries earlier, or perhaps Michael and his colleagues picked it up in the Middle East on one of their trade voyages.

One of Michael's procedures, simplified here, works something like the following. Suppose he wishes to sail 100 km eastward from X to his destination Y (Figure 42). However, a storm blows him 40° south of east some unknown distance, and he ends up at point Z. Fortunately, there's a lighthouse at Y with a fire at the top. Michael changes his direction and heads towards the lighthouse. Eventually he arrives at Y, forming an angle of 60° with his original eastward path. The question is: how far did Michael sail when he travelled from X to Z?

Since the angles in a triangle sum to 180°, we know immediately that the angle at Z is 80°. Michael draws a line across the sea from X so that it meets YZ at a right angle at A. Since XYA is a right triangle, we know that

$$\sin 60° = \frac{XA}{XY}.$$

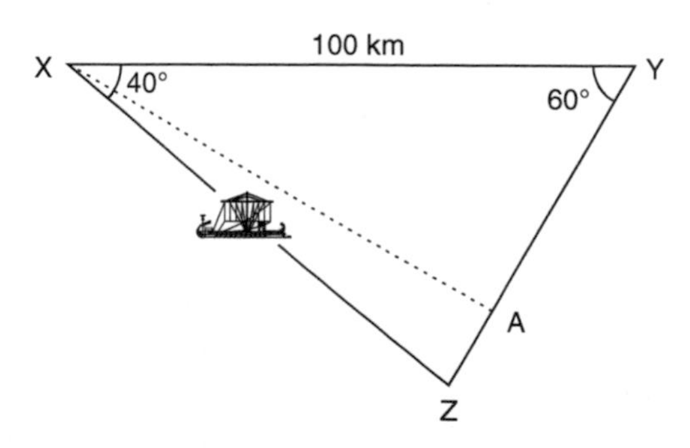

42. Michael of Rhodes' *marteloio* problem.

But XAZ is also a right triangle, so we also have

$$\sin 80° = \frac{XA}{XZ}.$$

If we solve both of these equations for XA and set them equal to each other, we end up with

$$\frac{XZ}{\sin 60°} = \frac{XY}{\sin 80°}.$$

This expression refers only to parts of the original triangle, not to the line XA that Michael had drawn across it. And there was nothing special about this triangle and the sides and angles we chose to work with. So, for any triangle whatever with sides a, b, and c, and angles A, B, and C opposite them, we have a new identity:

$$\text{LAW OF SINES: } \frac{a}{\sin A} = \frac{b}{\sin B} = \frac{c}{\sin C}$$

What makes this new identity so powerful is that it applies to any triangle at all. We no longer have to seek out right triangles, or break up our given triangle into right triangles as Michael did above. We can now solve Michael's problem easily enough. Since $XY = 100$ km, we have

$$\frac{XZ}{\sin 60°} = \frac{100}{\sin 80°};$$

thus $XZ = 100 \cdot \frac{\sin 60°}{\sin 80°} = 87.94$ km.

However, it's a dangerous business to assert from this evidence that Michael knew the Law of Sines. Faced with similar situations, Michael always divided his triangle into two right triangles.

For instance, in this case he would solve $\sin 60° = \frac{XA}{XY}$ to get $XA = 86.60$ km, and then solve $\sin 80° = \frac{XA}{XZ}$ to get $XZ = 87.94$ km. He never recognized the Law of Sines as a single entity, the way that we write it. On the other hand, clearly he could solve every problem that we can solve using the Law of Sines. So, does Michael deserve credit for the Law of Sines or not? You decide.

There's a problem with the Law of Sines. Suppose we know that $XY = 100$ km, $XZ = 87.94$ km, and the angle at Y is 60°. We can use the Law of Sines to calculate the angle at Z easily enough:

$$\frac{87.94}{\sin 60°} = \frac{100}{\sin Z},$$

which gives $\sin Z = 0.9848$, so (using the inverse sine button on our calculator) $Z = 80°$. But not so fast: if you evaluate $\sin 100°$, you *also* get 0.9848. So there are two possibilities for angle Z, and Figure 43 shows that both values lead to a plausible situation for Michael's ship. This ambiguity arises frequently when using the Law of Sines to determine an angle. In a navigational context, such ambiguities could put a ship in jeopardy.

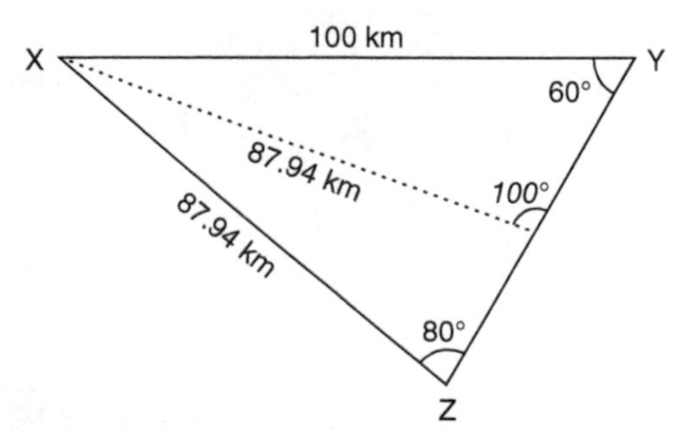

43. An ambiguity when using the Law of Sines.

The method we used to solve Michael of Rhodes' problem is identical to techniques used by surveyors today. Imagine a flat landscape divided into triangles of various shapes and sizes. If we know the length of one side of one triangle and the angles at each end of that side, the Law of Sines gives us the remaining dimensions of that triangle. Then we can go on to find the dimensions of an adjoining triangle, and so on. This *triangulation* was used in the last decade of the 18th century by Jean-Baptiste-Joseph Delambre and Pierre Méchain to calculate the north–south distance from Dunkerque to Barcelona through Paris. Their goal was to determine the length of the metre, which had been defined to be one ten millionth of the distance from the North Pole to the equator through Paris. Of course, since the ground is never perfectly flat, the problem is a bit more complicated than we have portrayed it here (see Figure 44).

44. The 'trig station', or survey marker, atop University of Canterbury Mount John Observatory, Lake Tekapo, New Zealand, the southernmost optical research facility in the world.

The Law of Cosines

Knowing where you are can be a matter of life and death. Consider the following situation (again, simplified): you are lost in the wilderness, with only your smartphone and a supply of trail mix to fend off the hunger. Although the GPS in your phone (at *A* in Figure 45) is not working properly, the phone itself is able to make contact with two cell phone towers whose positions *B* and *C* are known, and where your rescuers are waiting. The stations are 11,000 metres apart, and from the time it takes your phone to get signals from them, you know that you are 9,500 metres away from station *B* and 3,500 metres away from station *C*.

Triangle *ABC* isn't right angled, but if we drop a vertical line downward from *A*, we've broken it into two right triangles. As the sides are labelled in Figure 45, applying the Pythagorean theorem to the triangle on the left gets us

$$c^2 = (a-g)^2 + h^2 = a^2 - 2ag + g^2 + h^2.$$

But from the triangle on the right, we know that $g^2 + h^2 = b^2$. Hence

$$c^2 = a^2 + b^2 - 2ag.$$

Again from the triangle on the right, we know that $\cos C = \frac{g}{b}$, or $g = b\cos C$. Substituting into the above, we have the

$$\text{LAW OF COSINES}: c^2 = a^2 + b^2 - 2ab\cos C.$$

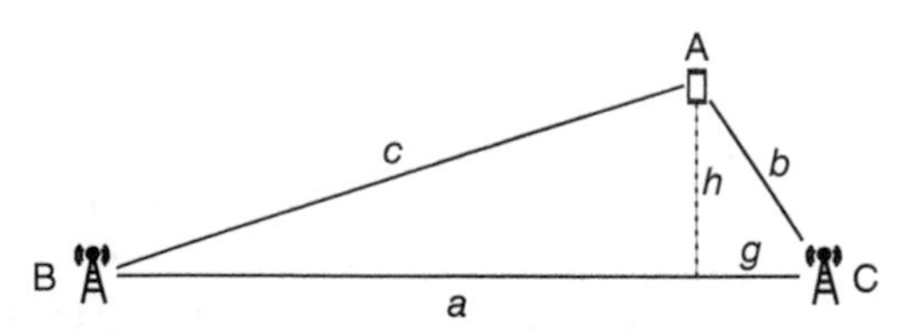

45. The Law of Cosines.

By the way, notice that the Pythagorean theorem is hidden here in plain sight: if angle C is 90°, then $\cos C = 0$ and we're left with $c^2 = a^2 + b^2$.

Returning to our cell phone situation, we know that $a = 11{,}000$ metres, $b = 3{,}500$ metres, and $c = 9{,}500$ metres. Substituting these values into the Law of Cosines, we find $C = 56.05°$. We ask the rescuers at C to head 56° to the right of the line connecting the stations, and before long we'll be rescued.

The derivation given here is similar to the one found in Book II of Euclid's classic geometry textbook, the *Elements*—which is surprising and a bit disturbing, since the *Elements* was written at least a century before trigonometry was invented and long before cosines were. But Euclid demonstrated simply that $c^2 = a^2 + b^2 - 2ag$; and he never used it in the way that we did, by substituting numerical values into an equation. It is nevertheless a curious fact that a geometric equivalent of the Law of Cosines was known well before trigonometry was invented.

Mollweide's formulas

The Laws of Sines and Cosines may be familiar, but here is a pair of formulas that few people see today. In any triangle where A, B, and C are the vertices and a, b, and c are the sides opposite the respective angles, then

$$\frac{a-b}{c} = \frac{\sin\frac{1}{2}(A-B)}{\cos\frac{1}{2}C} \quad \text{and} \quad \frac{a+b}{c} = \frac{\cos\frac{1}{2}(A-B)}{\sin\frac{1}{2}C}.$$

At first glance it appears that these formulas are not at all helpful. To use them you need to know five of the six elements of a triangle to determine the sixth. But if you know five of the six elements, we already have the Laws of Sines and Cosines to find the sixth. However, the very fact that the formulas contain all six elements

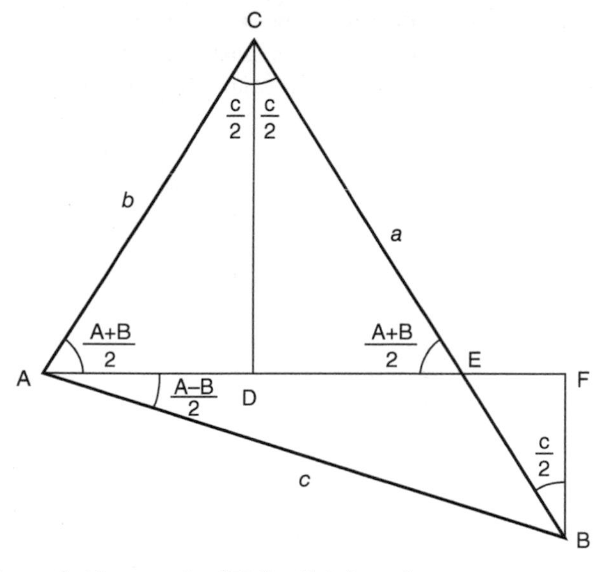

46. A proof without words of Mollweide's formula.

makes them handy: once you have completely solved a triangle, Mollweide provides a quick check to make sure that all your six values are correct.

The formulas are named for early 18th-century German astronomer Karl Mollweide (most famous today for the map projection named after him), although in Mollweide's 1808 paper he quotes a 1786 publication by Antonio Cagnoli in which they also appear. The first of the two formulas may be proved using Figure 46. We leave it as a challenge to see why.

A few tangents

In the past chapter and a half, all the identities we've explored have involved sines and cosines, but never tangents or other functions. That's not because there aren't any; rather, since the sine and cosine are the building blocks for trigonometry, the identities for sines and cosines are usually the first ones to which we turn. But tangents and the other functions deserve their time in the Sun, so let's take a look.

Our first and most well-known tangent identity is obtained by dividing the first of Mollweide's formulas by the second:

$$\frac{a-b}{a+b}=\frac{\tan\frac{1}{2}(A-B)}{\cot\frac{1}{2}C},$$

and then noticing that $\frac{1}{2}C=\frac{1}{2}(180°-A-B)=90°-\frac{A+B}{2}$. This gives us the

$$\text{LAW OF TANGENTS: }\frac{a-b}{a+b}=\frac{\tan\frac{1}{2}(A-B)}{\tan\frac{1}{2}(A+B)}.$$

Since we can already solve triangles with the Laws of Sines and Cosines, this new identity doesn't give us any more power than we had before. However, before we invented machines to do calculations for us, there were certain circumstances where the Law of Tangents was easier, for instance, in cases when the Law of Cosines requires the calculation of a square root to find a side length.

There is also a Pythagorean theorem of sorts for the tangent functions. Simply divide $\sin^2\theta+\cos^2\theta=1$ through by $\cos^2\theta$, and we have

$$\tan^2\theta+1=\sec^2\theta.$$

This identity was used occasionally to calculate values in secant tables; in certain circumstances it is less prone to error than the usual $\sec\theta=1/\cos\theta$.

There are double- and triple-angle formulas for the tangent, just as there are for the sine and cosine:

TANGENT DOUBLE-ANGLE FORMULA: $\tan 2\theta = \dfrac{2\tan\theta}{1-\tan^2\theta}$

TANGENT TRIPLE-ANGLE FORMULA: $\tan 3\theta = \dfrac{3\tan\theta - \tan^3\theta}{1-3\tan^2\theta}$

as well as similar formulas for the other functions; for instance,

COSECANT TRIPLE-ANGLE FORMULA:

$$\operatorname{cosec} 3\theta = \frac{\operatorname{cosec}^3\theta}{3\operatorname{cosec}^2\theta - 4}.$$

It will come, then, as no surprise that there are also sum and difference laws for the tangent and other functions. For instance:

TANGENT SUM AND DIFFERENCE LAWS:

$$\tan(\alpha \pm \beta) = \frac{\tan\alpha \pm \tan\beta}{1 \mp \tan\alpha\tan\beta}.$$

Many of these identities are very symmetric:

$$\operatorname{cosec}(\alpha+\beta+\gamma) = \frac{\sec\alpha\sec\beta\sec\gamma}{\tan\alpha+\tan\beta+\tan\gamma-\tan\alpha\tan\beta\tan\gamma}.$$

The list is endless; for any combination of functions, sums, and differences that you care to choose, there is likely an identity for it.

We have been avoiding any talk of identities for inverse trigonometric functions, partly because there aren't many of them. Several that involve the inverse tangent have surprising applications. However, since they arise naturally in Chapter 5, we shall keep you in suspense until then.

Products to sums, sums to products

Let's take a closer look at the sine sum and difference laws from Chapter 3:

$$\sin(\alpha+\beta)=\sin\alpha\cos\beta+\cos\alpha\sin\beta$$
$$\sin(\alpha-\beta)=\sin\alpha\cos\beta-\cos\alpha\sin\beta$$

The two expressions share the term $\cos\alpha\sin\beta$ on the right side: one added, the other subtracted. If we add the two equations together, those terms will cancel:

$$\sin(\alpha+\beta)+\sin(\alpha-\beta)=2\sin\alpha\cos\beta.$$

Solve this for $\sin\alpha\cos\beta$, and we get

$$\sin\alpha\cos\beta=\frac{1}{2}\left(\sin(\alpha+\beta)+\sin(\alpha-\beta)\right).$$

Similar manoeuvrings lead to the following:

$$\text{THE PRODUCT–TO–SUM FORMULAS:}$$
$$\sin\alpha\cos\beta=\frac{1}{2}\left(\sin(\alpha+\beta)+\sin(\alpha-\beta)\right)$$
$$\cos\alpha\sin\beta=\frac{1}{2}\left(\sin(\alpha+\beta)-\sin(\alpha-\beta)\right)$$
$$\sin\alpha\sin\beta=\frac{1}{2}\left(\cos(\alpha-\beta)-\cos(\alpha+\beta)\right)$$
$$\cos\alpha\cos\beta=\frac{1}{2}\left(\cos(\alpha+\beta)+\cos(\alpha-\beta)\right)$$

It may be unclear to us why we should bother with these formulas. In the 16th century they turned out to be a life saver. First discovered by Johann Werner, a Catholic priest in early 16th-century Nuremberg, they became part of daily astronomical life in Tycho Brahe's observatory on the island of Hven just north of Copenhagen. The huge astronomical instruments that Brahe built there were able to record observations of unprecedented accuracy;

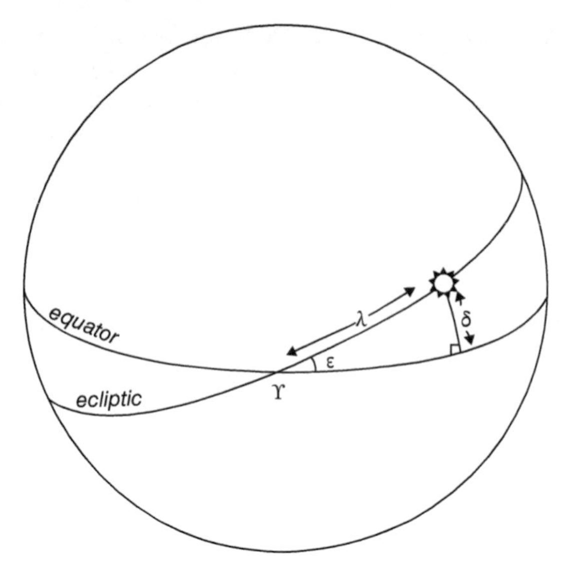

47. The solar declination.

they were eventually used by his assistant Johannes Kepler to help demonstrate that the planets move in ellipses rather than circles.

To see why the product-to-sum formulas were so valuable to Brahe's group, we borrow from Chapter 7 one of the most fundamental formulas from spherical astronomy. On the celestial sphere (Figure 47) we find the *celestial equator* and the *ecliptic*, the Sun's path through the celestial sphere on which the Sun travels about 1° per day. The Sun crosses the *vernal equinox* ♈ around March 21, the beginning of spring in the northern hemisphere. The angle between the two circles is $\varepsilon \approx 23.44°$, equivalent to the tilt of the Earth's axis. Given the time of year, we know the distance λ of the Sun from ♈. We wish to calculate the Sun's distance δ to the equator, its *declination*, using this formula:

$$\sin\delta = \sin\lambda \sin\varepsilon.$$

On May 20, λ is about 59°. To find δ, we must first multiply $\sin 59°$ by $\sin 23.44°$, which is 0.857167×0.397789. Today we simply reach for our calculators, hardly giving the multiplication a second thought. But place yourself in the shoes of Brahe's assistant, and put the calculator aside. Of course it's possible to multiply these numbers by hand, but it is both tedious and prone to error. Instead, we turn to the third product-to-sum formula to ease our computational woe:

$$\sin\delta = \frac{1}{2}\left(\cos(\lambda-\varepsilon)-\cos(\lambda+\varepsilon)\right) = \frac{1}{2}\left(\cos(35.56°)-\cos(82.44°)\right).$$

The product has disappeared, replaced by a much easier subtraction. Either way we come up with the answer $\delta = 19.94°$, but this new method (known as *prosthaphairesis*, from the Greek words for addition and subtraction) is more reliable and allows the astronomer's assistant to set aside the lengthy multiplications in favour of a good day's sleep.

It was this situation that led John Napier, only a couple of decades later, to invent logarithms. This is precisely what logarithms were designed to do: they turn products into sums, using the standard formula $\log xy = \log x + \log y$. If we apply it to our declination formula, we get

$$\log\sin\delta = \log\sin\lambda + \log\sin\varepsilon,$$

and the irksome product has vanished. The 'log sin', ugly as it looks, is not a problem: Napier and others provided tables of this combined function so users could simply look up their values.

Very quickly, logarithms took over from prosthaphairesis as the preferred computing tool in mathematical astronomy. Then, within a few years, they started to spread into everyday life. In fact logarithms can be credited, at least in part, for the rise of the use of mathematics in practical disciplines like surveying and architecture in the early 17th century, helping to lead eventually to modern science and technology.

For our next algebraic feat, let's turn the product-to-sum identities backwards. In the first product-to-sum identity above, set $x = \alpha + \beta$ and $y = \alpha - \beta$. A little algebra almost immediately gives

$$\sin x + \sin y = 2\sin\frac{x+y}{2}\cos\frac{x-y}{2}.$$

Performing similar tricks on all the product-to-sum identities gives us the following list:

THE SUM-TO-PRODUCT FORMULAS:

$$\cos x + \cos y = 2\cos\frac{x+y}{2}\cos\frac{x-y}{2}$$

$$\cos x - \cos y = -2\sin\frac{x+y}{2}\sin\frac{x-y}{2}$$

$$\sin x + \sin y = 2\sin\frac{x+y}{2}\cos\frac{x-y}{2}$$

$$\sin x - \cos y = 2\cos\frac{x+y}{2}\sin\frac{x-y}{2}$$

Clearly Brahe's assistants would have stayed away from these formulas like the plague. Why make life harder by turning a sum into a product? But there's a little bit of good, it seems, in everything. These apparently perverse formulas provide enlightenment in a rather surprising topic: the mathematics of music.

A pure musical tone is a sound wave that can be represented by the expression

$$c\sin(k \cdot 2\pi t),$$

where c is the amplitude of the sound, k is the frequency, and t is the time elapsed in seconds. The base note on the piano keyboard,

A above middle C, is set to 440 beats per second, or Hertz (Hz). Suppose that your piano's string, unbeknownst to you, is resonating at 444 Hz. You hire a piano tuner, who plays her reference note at the same time that she plays the note on your piano. The sound of the two pitches, close to each other but different, is discordant. You notice a pulsating increase and decrease in volume, a 'wah-wah-wah' sound known as the *beat phenomenon*.

Your piano tuner is well aware of this effect. When the notes are played simultaneously, our ears hear the sound added together:

$$\sin(444 \cdot 2\pi t) + \sin(440 \cdot 2\pi t).$$

This expression is in the form of our third sum-to-product formula. When we apply it, our expression transforms to

$$2\sin(442 \cdot 2\pi t)\cos(2 \cdot 2\pi t)$$

The $\sin(442 \cdot 2\pi t)$ term, the average of our two notes at 442 Hz, is the dominant frequency that we hear. But this sine wave is multiplied by $\cos(2 \cdot 2\pi t)$. From the graph in Figure 48, we can see that this causes the 442 Hz wave to pulsate, with a frequency of four beats per second (an 'up' and 'down' part of the wave for both of the cycles). So, when you hear the 'wah-wah-wah' sound, you are hearing the sum-to-product formula at work! Mindful of this phenomenon, the piano tuner knows that your A string is off by four beats per second. She is able to adjust your string accordingly, and restore your piano to perfect pitch.

48. The graph of $\sin(444 \cdot 2\pi t) + \sin(440 \cdot 2\pi t)$, illustrating the beat phenomenon. From the product-to-sum formula, this can also be written as $2\sin(442 \cdot 2\pi t)\cos(2 \cdot 2\pi t)$. The cosine term, graphed as a dashed curve, causes the amplitude of the 442 Hz wave to fluctuate periodically.

Morrie's Law and friends

Here's a mathematical magic trick: using your calculator, find $\cos 20°$. As expected, it's a complicated irrational number. Multiply it by $\cos 40°$. Again, a string of apparently random digits. Finally, multiply the result by $\cos 80°$. What do you get?

This rather curious result was shown to physicist Richard Feynman by his childhood friend Morrie Jacobs. Feynman never forgot it, and one can see why. Multiplying three seemingly unrelated irrational numbers and ending up with a simple fraction (for readers without a calculator, you get 0.125 or 1/8) demands an explanation; this sort of thing just doesn't happen by chance. Our journey into the reason why Morrie's Law works will also give us a fringe benefit: a chance to contrast how algebra and geometry can provide us with two different perspectives on the same situation.

We start with algebra. Recall, from Chapter 2, the sine angle sum formula:

SINE ANGLE SUM FORMULA:

$$\sin(\alpha+\beta)=\sin\alpha\cos\beta+\cos\alpha\sin\beta.$$

If we let $\beta=\alpha$, we get something useful:

SINE DOUBLE-ANGLE FORMULA: $\sin(2\alpha)=2\sin\alpha\cos\alpha.$

Morrie's Law is all about cosines, so we rearrange this to get

$$\cos\alpha=\frac{\sin 2\alpha}{2\sin\alpha}.$$

We can use our new formula to replace all the cosines in Morrie's Law with sines:

$$\cos 20°\cdot\cos 40°\cdot\cos 80°=\frac{\sin 40°}{2\sin 20°}\cdot\frac{\sin 80°}{2\sin 40°}\cdot\frac{\sin 160°}{2\sin 80°}.$$

Conveniently, the $\sin 40°$ and $\sin 80°$ terms cancel, leaving us with

$$\frac{\sin 160°}{8\sin 20°}.$$

But since 20° and 160° sum to 180°, their sines are equal, and they also cancel! We are left with the wonderful result

$$\cos 20° \cdot \cos 40° \cdot \cos 80° = \frac{1}{8}.$$

There are more nuggets to be found here, so let's keep digging. If we hadn't substituted 20° for α, we would have ended up with

$$\cos\alpha \cdot \cos 2\alpha \cdot \cos 4\alpha = \frac{\sin 2\alpha}{2\sin\alpha} \cdot \frac{\sin 4\alpha}{2\sin 2\alpha} \cdot \frac{\sin 8\alpha}{2\sin 4\alpha} = \frac{\sin 8\alpha}{8\sin\alpha}.$$

We didn't have write a product of exactly three cosines for the 'telescoping' cancellations to occur. We could write only two cosines, or four or five, or as many as we like. For instance, if we choose only two cosines, we get

$$\cos\alpha \cdot \cos 2\alpha = \frac{\sin 4\alpha}{4\sin\alpha}.$$

What made Morrie's Law work out so nicely was that the two sines on the right side of the equal sign cancelled out, and this happens when 4α is equal to $180° - \alpha$. That makes $\alpha = 36°$; pop that in and we get the satisfying

$$\cos 36° \cdot \cos 72° = \frac{1}{4}.$$

If we apply the same process with four cosines rather than two we also get a result, although the angle values aren't as pretty this time:

$$\cos\frac{180°}{17} \cdot \cos\frac{2 \cdot 180°}{17} \cdot \cos\frac{4 \cdot 180°}{17} \cdot \cos\frac{8 \cdot 180°}{17} = \frac{1}{16}.$$

There is another trick we can use for the four cosines. Up to now we have been cancelling the sines on the right side of the equal sign by choosing α so that the angle in the numerator is 180° minus the angle in the denominator. But we can also choose α so that the angle in the numerator is 180° *plus* the angle in the denominator. In this case the sines almost cancel, leaving behind −1. (See the graphs in Chapter 2 to recall why this is true.) We get

$$\cos 12° \cdot \cos 24° \cdot \cos 48° \cdot \cos 96° = -\frac{1}{16}.$$

There is more to find here, but at this point we pass the baton to the interested reader.

Let's change our point of view from algebra to geometry. We begin with the equation

$$\cos 36° \cdot \cos 72° = \frac{1}{4}.$$

In Figure 49 the regular pentagon has side lengths equal to 1. The angles in a regular pentagon are 108°, from which one can work out the angles indicated in the diagram. In triangle ABE we know that $AE = \cos 36°$, hence $AC = 2\cos 36°$. Then, from triangle ACF, $\cos 72° = CF / AC$, so $CF = 2\cos 36° \cdot \cos 72°$. But CF is half of the bottom side of the pentagon, so we already know that its length is ½. This demonstrates the equation.

For the original Morrie's Law, we move to a nonagon (Figure 50), again with side lengths equal to 1, and add an extra step to the process. The angles in a regular nonagon are 140°, and again we leave it to the reader to work out the other angles indicated in the diagram. From triangle ABF we know that $AF = \cos 20°$, hence $AC = 2\cos 20°$. Moving to the dashed lines and triangle AGC, we know that $AG = AC\cos 40° = 2\cos 20° \cos 40°$, so $AD = 4\cos 20° \cos 40°$. Finally, from triangle ADH, $DH = AD\cos 80° = 4\cos 20° \cos 40° \cos 80°$. But DH is half the

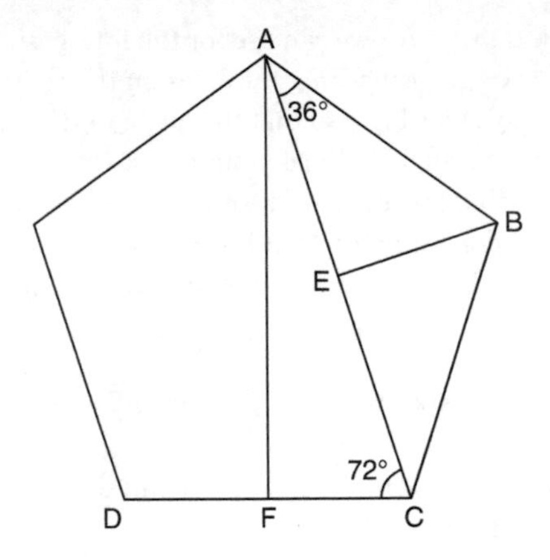

49. Morrie's pentagon.

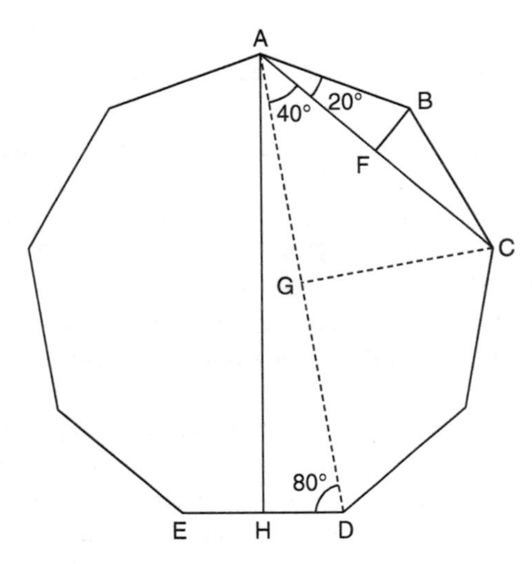

50. Morrie's nonagon.

bottom side of the nonagon, whose length we already know to be ½; and we have demonstrated Morrie's Law geometrically.

One can imagine extending this argument further and further (for the next step we need a 17-gon), proving the formula for more and more products of cosines—a repeating geometric algorithm, so to speak. The geometric and the algebraic arguments feel quite different, but they accomplish the same thing. Which one you prefer says something about your style of learning, so we leave you here to reflect on your mathematical personality.

The trigonometric zoo contains many more identities; we have seen only a taste here. Continuing this exploration will lead to many unexpected discoveries.

Chapter 5
To infinity...

The next few of pages will take us on a journey that goes on forever, and yet reaches an end. If you haven't seen infinity used in mathematics before, it might bother you. Some people get rather upset about it. When you get to a part of the argument that unsettles you, ask yourself precisely what you believe is wrong. We'll discuss these issues once we're done.

Consider the sector of a unit circle in Figure 51, with angle θ measured in radians. Since the area of the entire circle is $\pi r^2 = \pi \cdot 1^2 = \pi$, and we have the fraction $\theta / 2\pi$ of that circle, the area of our sector is $\pi \cdot \theta / 2\pi = \theta / 2$. But we're going to find the area in another, more difficult way. Eventually, we will combine our two area formulas together to discover something interesting.

We begin by breaking off the lightly shaded triangle; we label its area $T(\theta)$. Assuming the triangle's base to be the vertical side, we can see that

$$T(\theta) = \frac{1}{2}(base)(height) = \frac{1}{2}\left(2\sin\frac{\theta}{2}\right)\left(\cos\frac{\theta}{2}\right) = \sin\frac{\theta}{2}\cos\frac{\theta}{2}.$$

This looks a lot like the sine double-angle formula from Chapter 4. Using it, our expression simplifies to

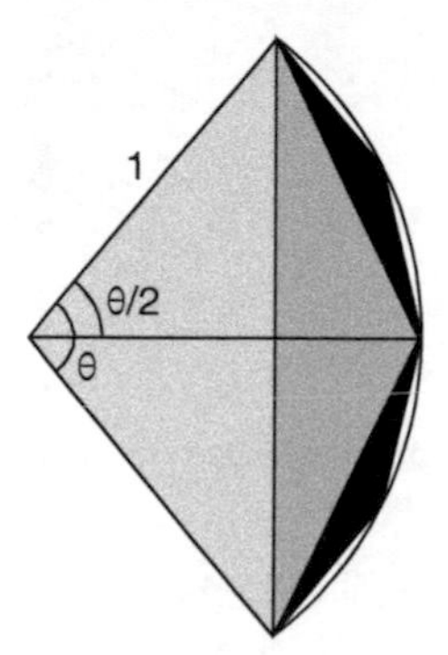

51. Deriving the Taylor series for the sine.

$$T(\theta)=\frac{1}{2}\sin\theta.$$

We turn next to the area of the triangle shaded dark grey, which we call $A(\theta)$. This time, see if you can verify the formula:

$$A(\theta)=\sin\frac{\theta}{2}\left(1-\cos\frac{\theta}{2}\right).$$

The expression in the parentheses might look familiar; it is the versed sine from Chapter 2. It also looks very much like the sine half-angle formula from Chapter 3:

$$A(\theta)=\sin\frac{\theta}{2}\left(2\sin^2\frac{\theta}{4}\right).$$

Let's add a twist. Suppose that θ is small. Since we're measuring angles in radians, we recall from the end of Chapter 3 that the sine of any small angle is roughly equal to the angle itself. So

$$A(\theta)=\sin\frac{\theta}{2}\left(2\sin^2\frac{\theta}{4}\right)\approx\frac{\theta}{2}\cdot 2\left(\frac{\theta}{4}\right)^2=\frac{\theta^3}{16}. \qquad (*)$$

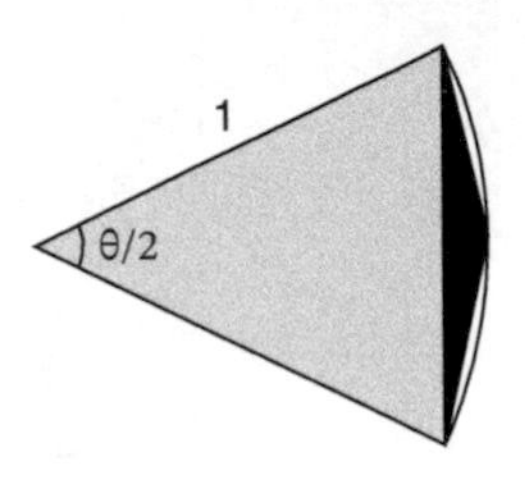

52. The top half of the previous figure, rotated clockwise.

Now, we want to find the area of the entire sector. But $T(\theta)$ and $A(\theta)$ do not fill it all in; there are two small sectors left out. We can get close to filling these sectors with the two black triangles in Figure 51, but what are the areas of these triangles? Consider Figure 52, which is the top half of Figure 51 rotated clockwise a little. The two figures appear almost identical, with two differences: the angle at the circle's centre is now $\frac{\theta}{2}$ rather than θ; and Figure 52 has a black triangle where Figure 51 had a triangle shaded dark grey. The area of each black triangle, then, can be found by repeating the calculation of $A(\theta)$ but starting instead with $\frac{\theta}{2}$; in other words, it is $A(\theta/2)$.

Four very small sectors remain unaccounted for. By the same reasoning, we can fill them with four triangles (too small to draw in our diagram), each with area $A(\theta/4)$. Now we have eight tiny sectors left over, which we fill with triangles of area $A(\theta/8)$. And so on, and so on, and so on. If we let the process continue *infinitely*, we will eventually (!) fill up the entire sector. So, the total area is

$$T(\theta)+A(\theta)+2A(\theta/2)+4A(\theta/4)+8A(\theta/8)+\ldots,$$

allowing the sum to go to infinity.

If we can swallow this, we're home free. We already know that the sector's area is $\theta/2$; we have a formula for $T(\theta)$; and we have a formula for $A(\theta)$ above (*) that is true for any value of θ. We bring it all together:

$$\begin{aligned}\text{Area} &= \frac{\theta}{2} = T(\theta) + A(\theta) + 2A(\theta/2) + 4A(\theta/4) + 8A(\theta/8) + \dots \\ &\approx \frac{1}{2}\sin\theta + \frac{\theta^3}{16} + 2\left(\frac{(\theta/2)^3}{16}\right) + 4\left(\frac{(\theta/4)^3}{16}\right) + 8\left(\frac{(\theta/8)^3}{16}\right) + \dots \\ &= \frac{1}{2}\sin\theta + \frac{\theta^3}{16} + \frac{\theta^3}{64} + \frac{\theta^3}{256} + \frac{\theta^3}{1{,}024} + \dots\end{aligned}$$

What are we to do with this infinite collection of fractions? We have a very clever trick up our sleeve to deal with it, although it is controversial. It provoked the longest flame war in the Internet's history back in the days of Compuserve, and more recently triggered wildfires on World of Warcraft and Ayn Rand message boards. The question that set things off: 'Is $0.9999999\ldots = 1$?' Not *close* to 1, but *exactly* 1? The answer is 'yes', and here's why. Write $0.9999999\ldots$ in fractions:

$$\frac{9}{10} + \frac{9}{100} + \frac{9}{1{,}000} + \frac{9}{10{,}000} + \dots$$

Multiply this expression by $\frac{1}{10}$ and we get

$$\frac{9}{100} + \frac{9}{1{,}000} + \frac{9}{10{,}000} + \dots$$

The difference between these two expressions must be $\frac{9}{10}$ of $0.9999999\ldots$. But if you subtract the second sum of fractions from the first one, everything cancels out except the very first term, $\frac{9}{10}$! So

$$\frac{9}{10}(0.9999999\ldots) = \frac{9}{10},$$

and therefore $0.9999999\ldots = 1$.

We can do the same thing to our expression $\frac{\theta^3}{16}+\frac{\theta^3}{64}+\frac{\theta^3}{256}+\frac{\theta^3}{1{,}024}+\cdots$ (*hint*: rather than multiplying the expression by $\frac{1}{10}$, multiply it by $\frac{1}{4}$). When we do, it works out to be equal to $\theta^3/12$. Therefore, our sector's area is approximately

$$\frac{1}{2}\sin\theta+\frac{\theta^3}{12}.$$

But recall from the beginning of this chapter that the area is exactly equal to $\frac{\theta}{2}$. Therefore $\frac{\theta}{2}\approx\frac{1}{2}\sin\theta+\frac{\theta^3}{12}$, and solving for $\sin\theta$, we end up with

$$\sin\theta\approx\theta-\frac{\theta^3}{6}.$$

OK, let's sum up (no pun intended). We started out assuming that if θ is small, then $\sin\theta\approx\theta$. We now have a better approximation: if θ is small, then $\sin\theta$ is even closer to $\theta-\frac{\theta^3}{6}$. Here's a bright idea: let's go back to the first time we used $\sin\theta\approx\theta$ (our equation (*)), replace it with our improved $\sin\theta\approx\theta-\frac{\theta^3}{6}$, and see what we get. Although the algebra is messier, the process is the same as before; we get

$$\sin\theta\approx\theta-\frac{\theta^3}{6}+\frac{\theta^5}{120}.$$

No reason to stop now: take our even better approximation back to (*), and repeat. This time we get

$$\sin\theta\approx\theta-\frac{\theta^3}{6}+\frac{\theta^5}{120}-\frac{\theta^7}{5{,}040}.$$

Why not keep going? In fact, why not keep going *infinitely*? If we do, we get

$$\sin\theta = \theta - \frac{\theta^3}{6} + \frac{\theta^5}{120} - \frac{\theta^7}{5{,}040} + \frac{\theta^9}{362{,}880} - \frac{\theta^{11}}{39{,}916{,}800} + \frac{\theta^{13}}{6{,}227{,}020{,}800} - \dots$$

Those numbers in the denominators look ugly, but they conceal a hidden pattern: $6 = 3 \cdot 2 \cdot 1$, then $120 = 5 \cdot 4 \cdot 3 \cdot 2 \cdot 1$, then $5{,}040 = 7 \cdot 6 \cdot 5 \cdot 4 \cdot 3 \cdot 2 \cdot 1$, and so on. These numbers can be written as factorials, and we have the

TAYLOR SERIES FOR THE SINE:

$$\sin\theta = \theta - \frac{\theta^3}{3!} + \frac{\theta^5}{5!} - \frac{\theta^7}{7!} + \frac{\theta^9}{9!} - \frac{\theta^{11}}{11!} + \frac{\theta^{13}}{13!} - \dots$$

If our casual manipulations of expressions with infinitely many terms doesn't bother you, then feel free to skip ahead a couple of paragraphs. For the rest of us, there is something awfully strange and possibly illegal about permitting these sums to exist at all. We seem to have shifted from thinking of infinity as a *process*, to infinity as a *completed reality*. Can one really do that? One can add 0.09 to 0.9, and then add 0.009 and so on, but one can never add infinitely many terms together. This is certainly true, but it confuses what one can calculate with what one can contemplate. As long as humans have been able to think, we have played with the implications of allowing an infinite process to be completed—actual infinity, as opposed to potential infinity. Permitting ourselves to think in this way led in part to the most powerful mathematical tool the human race has ever invented: calculus, of which this Taylor series is an example. Isaac Newton, one of its inventors, wandered down paths similar to ours in his classic work *Of Analysis by Equations of an Infinite Number of Terms* (Figure 53). It wasn't until more than a century after Newton that this reasoning was put on a firm logical foundation. In any case, we'd best hope that these infinite deductions are correct, because calculus is the mathematical tool behind most

[321]

OF

ANALYSIS

BY

Equations of an infinite Number of Terms.

1. *THE General Method, which I had devised some considerable Time ago, for measuring the Quantity of Curves, by Means of Series, infinite in the Number of Terms, is rather shortly explained, than accurately demonstrated in what follows.*

2. Let the Base AB of any Curve AD have BD for it's perpendicular Ordinate; and call AB $=x$, BD $=y$, and let a, b, c, &c. be given Quantities, and m and n whole Numbers. Then

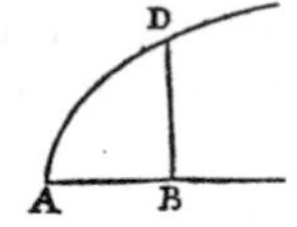

The Quadrature of Simple Curves,

RULE I.

3. If $ax^{\frac{m}{n}} = y$; it shall be $\frac{an}{m+n}x^{\frac{m+n}{n}}$ = Area ABD.

The thing will be evident by an Example.

1. If $x^2 (= 1x^{\frac{2}{1}}) = y$, that is $a = 1 = n$, and $m = 2$; it shall be $\frac{1}{3}x^3$ = ABD.

Tt 2. Suppose

53. The first page of Isaac Newton's *Of Analysis by Equations of an Infinite Number of Terms.*

of modern science and technology. Deny its validity, and risk your smartphone's demise.

If you're still not convinced (and there's no particular reason why you should be, at least yet), see Further Reading for sources that go into this more thoroughly. In the meantime, ask yourself whether you believe that $1/3 = 0.3333333\ldots$ Then multiply both sides of this equation by three.

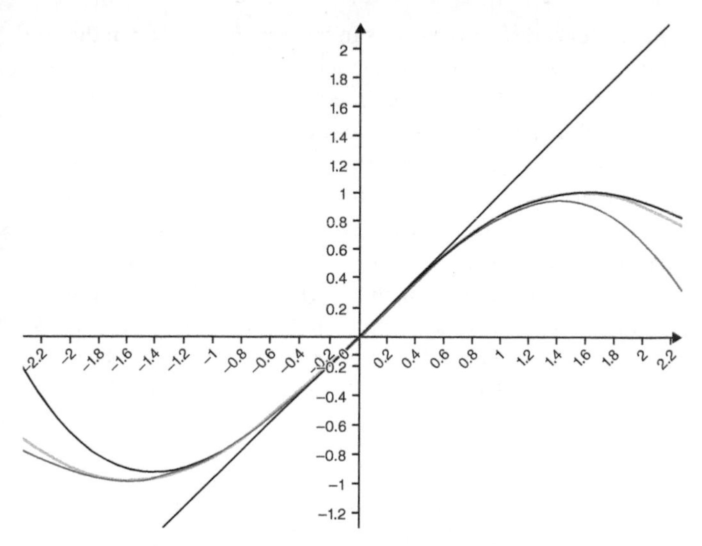

54. A graph of Taylor approximations to the sine curve. The grey curve is $y = \sin \theta$. The diagonal line is $y = \theta$; the thin curve that starts at the top left and ends at the bottom right is $y = \theta - \theta^3/6$; and the thin curve that starts at the bottom left and ends at the top right is $y = \theta - \theta^3/6 + \theta^5/120$.

We can see from a graph that the Taylor series works. Figure 54 shows the first several Taylor approximations, that is, $y = \theta$, $y = \theta - \theta^3 / 6$, and so on. The curves get closer and closer to the sine wave; and as we add more terms, the approximations fit the sine curve well for a wider and wider range of values of θ. Part of the reason that the Taylor series fits so well is that each term in the sequence is much smaller than the terms before it. The factorials in the denominators of the fractions become very large, very quickly.

If Isaac Newton was already using infinite series, one might wonder how they came to be named after someone else. The reason is that only some infinite series are named after Taylor. Scottish mathematician James Gregory (1638–75) worked with them, and Brook Taylor (1685–1731) eventually came up with a general method to construct them in 1715. That was enough to get him the name.

Our Taylor series for the sine is in fact much older. From the late 14th century to the early 16th, in Kerala in south-west India, an extraordinary collection of astronomers found it and several other series on their own—without the modern machinery of calculus. Beginning with Mādhava of Saṅgamagrāma, these scholars were searching for better ways to calculate sines and cosines, in order to build the best possible foundation for their astronomical work. They worked entirely geometrically as we did earlier, without needing to define concepts from calculus like the derivative. The Keralite mathematicians relied on infinite processes much like we did (see Further Reading), but they computed more than just the sine series. For instance, they found the

TAYLOR SERIES FOR THE COSINE:

$$\cos\theta = 1 - \frac{\theta^2}{2!} + \frac{\theta^4}{4!} - \frac{\theta^6}{6!} + \frac{\theta^8}{8!} - \frac{\theta^{10}}{10!} + \frac{\theta^{12}}{12!} - \ldots$$

They also found what has until recently been called the Gregory–Leibniz series for π:

$$\frac{\pi}{4} = 1 - \frac{1}{3} + \frac{1}{5} - \frac{1}{7} + \frac{1}{9} - \ldots$$

In a rare example of history being corrected, this expression is gradually becoming known under an expanded name: the Mādhava–Gregory–Leibniz series.

Using infinite trigonometric series to compute π

If we wanted to compute an accurate value for π, the Mādhava–Gregory–Leibniz series is a very bad way to go about it. Although you will get there in the end, in this case the end is a very, very long way away. Here is what happens if you try:

$$\frac{\pi}{4} \approx 1 - \frac{1}{3} \qquad \pi \approx 2.66666667$$

$$\frac{\pi}{4} \approx 1 - \frac{1}{3} + \frac{1}{5} \qquad \pi \approx 3.46666667$$

$$\frac{\pi}{4} \approx 1 - \frac{1}{3} + \frac{1}{5} - \frac{1}{7} \qquad \pi \approx 2.89523810$$

$$\frac{\pi}{4} \approx 1 - \frac{1}{3} + \frac{1}{5} - \frac{1}{7} + \frac{1}{9} \qquad \pi \approx 3.33968254$$

$$\vdots \qquad \vdots$$

$$\frac{\pi}{4} \approx 1 - \frac{1}{3} + \frac{1}{5} - \frac{1}{7} + \ldots + \frac{1}{101} \qquad \pi \approx 3.16119861$$

$$\vdots \qquad \vdots$$

$$\frac{\pi}{4} \approx 1 - \frac{1}{3} + \frac{1}{5} - \frac{1}{7} + \ldots + \frac{1}{1{,}001} \qquad \pi \approx 3.14358866$$

Even after the $\frac{1}{1{,}001}$ term, we have π to only two decimal places! This does not bode well. If you want π accurate to ten decimal places, you'll need around five billion terms. However, there are some clever ways to speed up the convergence, a couple of which were discovered in Kerala by Mādhava himself. The simplest is to subtract a correction term (or add it, if your last term was subtracted) when you grow weary. The first correction term Mādhava came up with is simply the reciprocal of the number of terms that you have combined so far: for instance, in the case of $\frac{\pi}{4} \approx 1 - \frac{1}{3} + \frac{1}{5} - \frac{1}{7} + \frac{1}{9}$, subtract $\frac{1}{5}$. Our π value improves from $\pi \approx 3.33968254$ to the much better 3.13968254. When the last term is $\frac{1}{101}$ our estimate becomes 3.14159077, and when the last term is $\frac{1}{1{,}001}$ we get 3.14159265, accurate to all eight places. Much better.

We can find π even more effectively using trigonometric series. We consider next the brilliant idea conceived by John Machin (1680–1751), Brook Taylor's mathematics tutor before he attended Cambridge University. Taylor once told Machin that a comment he had made in a conversation over coffee had inspired Taylor to discover the theorem underlying the series that bear his name. Machin's method was described by his friend William Jones in 1706, in a passage that is the first ever to use the symbol π to refer to 3.1415.... Jones was a good salesman:

> But the method of series... performs this [the calculation of π] with great facility, when compared with the intricate prolix ways of Archimedes [and other geometers]. Tho' some of them were said to have (in this case) set bounds to human improvements, and to have left nothing for posterity to boast of; but we see no reason why the indefatigible labor of our ancestors should restrain us to those limits, which by means of the modern geometry, are made so easy to surpass.

This 'modern geometry', relying heavily on computation, was inspired by calculus and would eventually be called analysis. We use the same word in the phrase that describes the mathematics of x and y coordinates, 'analytic geometry'.

Machin's idea was to use the series for the inverse tangent, which is surprisingly simple:

$$\text{INVERSE TANGENT SERIES: } \tan^{-1} x = x - \frac{x^3}{3} + \frac{x^5}{5} - \frac{x^7}{7} + \frac{x^9}{9} - \ldots$$

Now, we know that $\tan 45° = 1$, but since we are now measuring angles in radians, we have $\tan\frac{\pi}{4} = 1$. If we let $x = 1$ in the inverse tangent series, we get just the Mādhava–Gregory–Leibniz series. But Machin had something shrewder in mind. In the unit circle of Figure 55, the heights of the right triangles are the tangents of their respective angles (because tangent = opposite/adjacent, and the adjacent sides are all equal to 1). We begin by marking off a height of $\frac{1}{5}$, the very thick line in the figure. The corresponding angle θ is $\tan^{-1}\frac{1}{5} \approx 11.31°$.

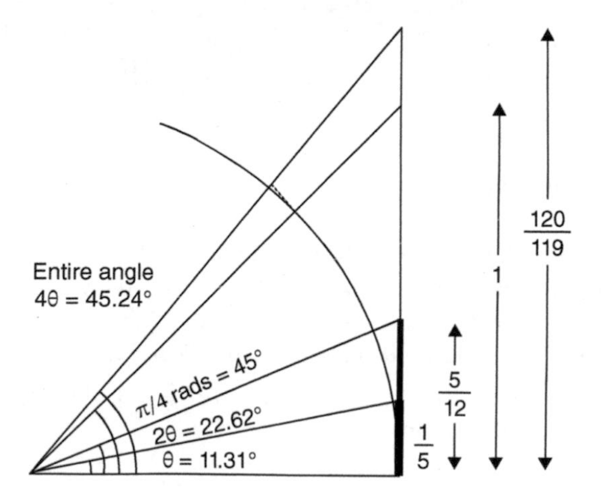

55. Machin's derivation of his inverse tangent formula. We have had to exaggerate the small angle by a factor of about twenty, in order to make the small dotted line barely visible.

It turns out that we can find the tangent of 2θ, the somewhat thick line, by recalling the tangent sum and difference laws from Chapter 4:

TANGENT SUM AND DIFFERENCE LAWS:

$$\tan(\alpha \pm \beta) = \frac{\tan\alpha \pm \tan\beta}{1 \mp \tan\alpha \tan\beta}.$$

If we set both α and β equal to θ and remember that $\tan\theta = \frac{1}{5}$, we find

$$\tan 2\theta = \frac{2\tan\theta}{1-\tan^2\theta} = \frac{2\cdot\frac{1}{5}}{1-\left(\frac{1}{5}\right)^2} = \frac{5}{12}.$$

So the height of the somewhat thick line is $\frac{5}{12}$. Let's double the angle again to 4θ. This gives us the height of the entire vertical line:

$$\tan 4\theta = \frac{2\tan 2\theta}{1-\tan^2 2\theta} = \frac{2\cdot\frac{5}{12}}{1-\left(\frac{5}{12}\right)^2} = \frac{120}{119}.$$

This value is very close to 1.

On its own this fact doesn't help us. But now consider the very short, barely visible diagonal dotted line that is tangent to the circle's 45° point, and traverses the gap to the highest diagonal line corresponding to 4θ. Since it's perpendicular to the radius at the 45° point, its length is equal to $\tan\left(4\theta - \frac{\pi}{4}\right)$. By the tangent difference law, its length is

$$\tan\left(4\theta - \frac{\pi}{4}\right) = \frac{\tan 4\theta - 1}{1+\tan 4\theta} = \frac{\frac{120}{119}-1}{1+\frac{120}{119}} = \frac{1}{239}.$$

We're almost there! From this, the very small angle $4\theta - \frac{\pi}{4}$ is equal to $\tan^{-1}\left(\frac{1}{239}\right)$. Substituting $\theta = \tan^{-1}\left(\frac{1}{5}\right)$ into this equation and solving for $\frac{\pi}{4}$, we have

$$\text{MACHIN'S FORMULA: } \frac{\pi}{4} = 4\tan^{-1}\left(\frac{1}{5}\right) - \tan^{-1}\left(\frac{1}{239}\right).$$

This formula has two very useful properties. Firstly, since we use a decimal system of numeration, when we plug $\frac{1}{5}$ into the inverse tangent series the calculations will be relatively easy:

$$\tan^{-1}\left(\frac{1}{5}\right) = \frac{1}{5} - \frac{\left(\frac{1}{5}\right)^3}{3} + \frac{\left(\frac{1}{5}\right)^5}{5} - \ldots = \frac{1}{5} - \frac{1}{375} + \frac{1}{15,625} - \ldots$$

When we plug in $\frac{1}{239}$, we lose that advantage—but $\frac{1}{239}$ is a very small quantity. So, secondly, the terms of the $\tan^{-1}\left(\frac{1}{239}\right)$ series get very small very quickly:

$$\tan^{-1}\left(\frac{1}{239}\right) = \frac{1}{239} - \frac{\left(\frac{1}{239}\right)^3}{3} + \frac{\left(\frac{1}{239}\right)^5}{5} - \dots$$
$$= \frac{1}{239} - \frac{1}{40{,}955{,}757} + \frac{1}{3{,}899{,}056{,}325{,}995} - \dots$$

Machin used his formula to compute π to 100 places, surely more accurate than was needed for any practical purpose. Successors quickly broke his record using similar techniques. Methods involving the inverse tangent series are still used even today by some modern π hunters.

How do calculators evaluate trigonometric functions?

By now I probably don't have to convince you that trigonometric functions can be challenging to work with, much more than ordinary arithmetic. We can easily figure out 46×34 on paper, but we needed most of Chapter 3 to determine $\sin 1°$. Taylor series allow us to convert the challenge of calculating a sine from a hard geometric problem to a merely tedious arithmetic one: just replace the sine with the Taylor series, and calculate for as many terms as you have patience. This is why Mādhava derived Taylor series in the first place.

When you ask calculus teachers how a calculator is able to evaluate sines, they are almost certain to respond 'Taylor series'. Evaluating a Taylor series involves nothing more than adding, subtracting, multiplying, and dividing numbers, which is what calculators were designed to do. But the teachers are wrong. Calculators do *not* use Taylor series. Rather, they use an entirely different algorithm, ingeniously designed to work even more smoothly and quickly.

In the late 1950s Jack Volder, a former B-24 flight engineer in World War II, was asked by his employer Corvair to replace the

B-58 bomber's analogue computer-driven navigation system with a digital equivalent. At the time, if you needed values of trigonometric functions, you still had to look them up in tables. Volder's solution, called CORDIC (*CO*ordinate *R*otation *DI*gital *C*omputer), was much faster than other methods. Variants of it are used by most pocket calculators today. Some of the ideas behind CORDIC go back to early 17th-century mathematician Henry Briggs, the inventor of base ten logarithms. In the early days of pocket calculators, Hewlett Packard was once served a notice of patent infringement from Wang Labs for using a similar algorithm. HP sent back a copy of Briggs's 17th-century Latin text, and that was the end of the matter.

Here's what happens when you type a number into your calculator, for example 67.89°, and press the sin key. First, the calculator breaks your number down into sums and differences of the following set of angles:

$$\theta_0 = \tan^{-1} 1 = 45^\circ$$

$$\theta_1 = \tan^{-1}\left(\frac{1}{2}\right) = 26.565^\circ$$

$$\theta_2 = \tan^{-1}\left(\frac{1}{4}\right) = 14.036^\circ$$

$$\theta_3 = \tan^{-1}\left(\frac{1}{8}\right) = 7.125^\circ$$

...and so on, up to around θ_{39}, which is on the order of 10^{-10} and just beyond the range of your calculator's accuracy. Each of these angles is roughly half of the preceding one. In our case,

$$\begin{aligned} 67.89^\circ &= 45^\circ + 26.565^\circ - 14.036^\circ + 7.125^\circ + \ldots - 0.0000000001042^\circ \\ &= \theta_0 + \theta_1 - \theta_2 + \theta_3 + \ldots - \theta_{39}. \end{aligned}$$

Figure 56 shows what's happening visually. On the unit circle, we seek the lengths $\cos 67.89^\circ$ and $\sin 67.89^\circ$, the two dashed lines

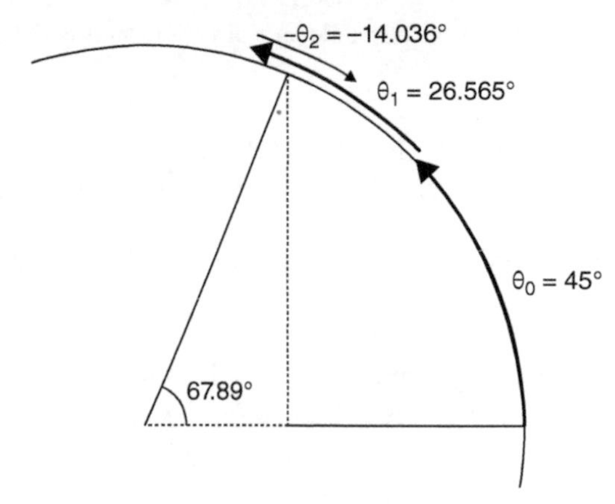

56. Using the CORDIC algorithm to calculate sin 67.89°. At each stage we either add or subtract the rotation θ_n, getting closer and closer to 67.89°. Only the first three of 40 steps are shown.

in the circle. We start by moving from the rightmost point on the circle (1,0) upward by $\theta_0 = 45°$, then upward again by $\theta_1 = 26.565°$, then downward by $\theta_2 = 14.036°$, and so on. At each step we rotate either upward or downward, whatever gets us closer to 67.89°. Recall from Chapter 3 that you can rotate a point around the origin by applying the following matrix:

$$\begin{bmatrix} \cos\theta & -\sin\theta \\ \sin\theta & \cos\theta \end{bmatrix}.$$

Our first rotation gives us

$$\begin{bmatrix} \cos\theta_0 & -\sin\theta_0 \\ \sin\theta_0 & \cos\theta_0 \end{bmatrix}\begin{bmatrix} 1 \\ 0 \end{bmatrix}.$$

Our second rotation applies a new matrix to this result, so we have

$$\begin{bmatrix} \cos\theta_1 & -\sin\theta_1 \\ \sin\theta_1 & \cos\theta_1 \end{bmatrix}\begin{bmatrix} \cos\theta_0 & -\sin\theta_0 \\ \sin\theta_0 & \cos\theta_0 \end{bmatrix}\begin{bmatrix} 1 \\ 0 \end{bmatrix}.$$

Our third rotation is in the backwards direction, so this time we have

$$\begin{bmatrix} \cos(-\theta_2) & -\sin(-\theta_2) \\ \sin(-\theta_2) & \cos(-\theta_2) \end{bmatrix} \begin{bmatrix} \cos\theta_1 & -\sin\theta_1 \\ \sin\theta_1 & \cos\theta_1 \end{bmatrix} \begin{bmatrix} \cos\theta_0 & -\sin\theta_0 \\ \sin\theta_0 & \cos\theta_0 \end{bmatrix} \begin{bmatrix} 1 \\ 0 \end{bmatrix}.$$

We repeat this process forty times, ending up with the horrendous expression

$$\begin{bmatrix} \cos\theta_{39} & \sin\theta_{39} \\ -\sin\theta_{39} & \cos\theta_{39} \end{bmatrix} \cdots \begin{bmatrix} \cos(\theta_2) & \sin(\theta_2) \\ -\sin(\theta_2) & \cos(\theta_2) \end{bmatrix}$$
$$\begin{bmatrix} \cos\theta_1 & -\sin\theta_1 \\ \sin\theta_1 & \cos\theta_1 \end{bmatrix} \begin{bmatrix} \cos\theta_0 & -\sin\theta_0 \\ \sin\theta_0 & \cos\theta_0 \end{bmatrix} \begin{bmatrix} 1 \\ 0 \end{bmatrix}.$$

This will generate the values of the sine and cosine of 67.89° accurate to about eleven decimal places, but it hardly seems efficient. However, this is where Volder's inspired choice of the values of θ_n pays off. Each rotation matrix can be factored as follows:

$$\begin{bmatrix} \cos\theta_n & -\sin\theta_n \\ \sin\theta_n & \cos\theta_n \end{bmatrix} = \cos\theta_n \begin{bmatrix} 1 & -\tan\theta_n \\ \tan\theta_n & 1 \end{bmatrix};$$

but we know from how we chose the θ_ns that $\tan\theta_n = 2^{-n}$. Also

$$\cos\theta_n = \frac{1}{\sec\theta_n} = \frac{1}{\sqrt{1+\tan^2\theta_n}} = \frac{1}{\sqrt{1+2^{-2n}}}.$$

The whole matrix can be written without any trigonometry at all:

$$\begin{bmatrix} \cos\theta_n & -\sin\theta_n \\ \sin\theta_n & \cos\theta_n \end{bmatrix} = \frac{1}{\sqrt{1+2^{-2n}}} \begin{bmatrix} 1 & -2^{-n} \\ 2^{-n} & 1 \end{bmatrix}.$$

Calculators use binary arithmetic; for them, multiplying by a power of 2 is just as easy as multiplying by a power of 10 is for us. But there's another advantage. The forty quantities $\frac{1}{\sqrt{1+2^{-2n}}}$ are identical, regardless of what number the user enters into the calculator. We can simply gather these quantities together in advance, multiply them, and hardwire the result into the algorithm so that we do not have to figure it out every time. This number turns out to be 0.607252935. The whole mess thus simplifies to

$$0.607252935\begin{bmatrix}1 & 2^{-39}\\ -2^{-39} & 1\end{bmatrix}\cdots\begin{bmatrix}1 & 2^{-2}\\ -2^{-2} & 1\end{bmatrix}\begin{bmatrix}1 & -2^{-1}\\ 2^{-1} & 1\end{bmatrix}\begin{bmatrix}1 & -2^{0}\\ 2^{0} & 1\end{bmatrix}\begin{bmatrix}1\\ 0\end{bmatrix}.$$

The operations in this expression, almost all of which involve simple powers of 2, are very fast. You won't have time to blink between the button press and the result. We have come a long way from Chapter 3, where we learned how to build sines from elementary geometry. Bare hands are no longer needed.

Fourier and Kelvin, tides and music

Back in Chapter 1, we left 19th-century scientist Sir William Thomson (later Lord Kelvin) puzzling over how to predict ocean tides. The physics underlying tides is well understood; tides are the result of gravitational interactions between the Earth, Moon, and Sun. But knowing this interplay does not help us to predict the tide at a particular location. Local effects such as the height of the sea floor and the shape of the coastline alter tides dramatically, which is why they differ at different coasts. Predicting the tides, then, becomes a problem in reverse engineering: how can we use data about recent tides in some location to determine how the tides will behave there in the future?

Thomson had an advantage. At sixteen, his family went on an excursion to Germany, and his father banned all work so that they could practice their German. Teenaged Thomson snuck into his luggage his favourite book, and during their stay he hid in the cellar reading it. The author was not a 19th-century equivalent of J. K. Rowling, but rather Jean-Baptiste Joseph Fourier (1768–1830).

Earlier in the century, Fourier had solved the problem of how heat travels through a medium in his *Analytical Theory of Heat*; later he would be the first to argue for the existence of the greenhouse effect.

As we saw earlier in this chapter, Brook Taylor and his colleagues had tamed many functions by representing them as infinite sums of powers of x. Fourier turned the tables on Taylor: rather than building trigonometric functions using polynomials, he built a variety of functions using sines and cosines. Recall from Chapter 4 that we generate the sound of two pure tones played together by adding two sine functions vibrating at the notes' frequencies. For example, when a note of 444 Hz is played simultaneously and at the same volume as a note of 440 Hz, we get

$$\sin(444 \cdot 2\pi t) + \sin(440 \cdot 2\pi t).$$

Fourier discovered how to work in the other direction: from a given periodic function, he was able to break it down into a sum of sines and cosines, thereby decomposing the function into component 'notes'. For example, consider the *square wave* (Figure 57), used in digital switching circuits and to synthesize the sounds of wind instruments. Fourier's series for the square wave with an amplitude of 1 and period 2 is

$$\frac{4}{\pi}\sin(\pi t) + \frac{4}{3\pi}\sin(3\pi t) + \frac{4}{5\pi}\sin(5\pi t) + \frac{4}{7\pi}\sin(7\pi t) + \ldots$$

Fourier's solution was controversial. As successive sinusoidal terms are added, the curve gets closer and closer to the square wave. The controversy arises when one considers some of the strange things that happen to the series for certain values of t. In our example, although the curves get closer to the square wave as more terms are added, we can see on the graph that each Fourier curve 'overshoots' the square wave around the places

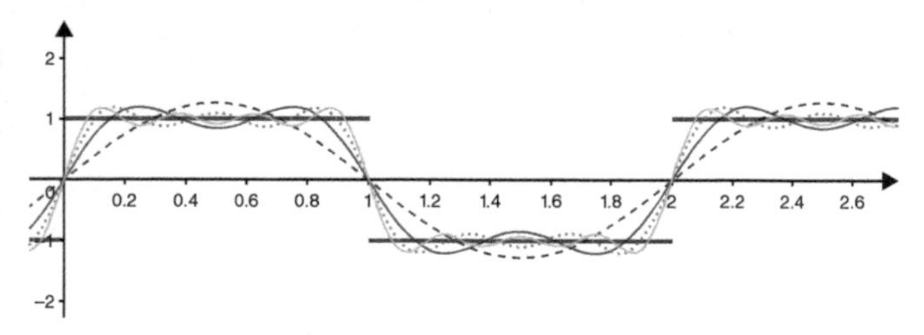

57. The square wave (drawn in thick lines) and the first four Fourier approximations to it. At their peaks, the approximating curves are all about 9 per cent above the square wave.

where the square wave jumps. Although the problem intervals get narrower and narrower as more terms are included, the heights of the overshoots remain around 9 per cent above and below the square wave. This effect is called the *Gibbs phenomenon.*

Now, Thomson's teenaged cellar reading had familiarized him with Fourier analysis, and he recognized immediately that the patterns of the tides also might be broken into a sum of sine waves. From the relative motions of the Sun and the Moon, he knew the frequencies v_n of the individual tidal components, but not their magnitudes. So his model of the tidal height as a function of time was

$$f(t) = A_0 + A_1 \sin(v_1 t) + B_1 \cos(v_1 t) + A_2 \sin(v_2 t) + B_2 \cos(v_2 t) + \dots.$$

Thomson's task was determine the values of the As and Bs for a given location from historical records. There are over a hundred frequencies v_n, but most analyses use only the largest twenty-five of them.

For our example we will keep things simple, using only two frequencies and only the sines:

$$f(t) = A_0 + A_1 \sin(t) + A_2 \sin(1.5678t).$$

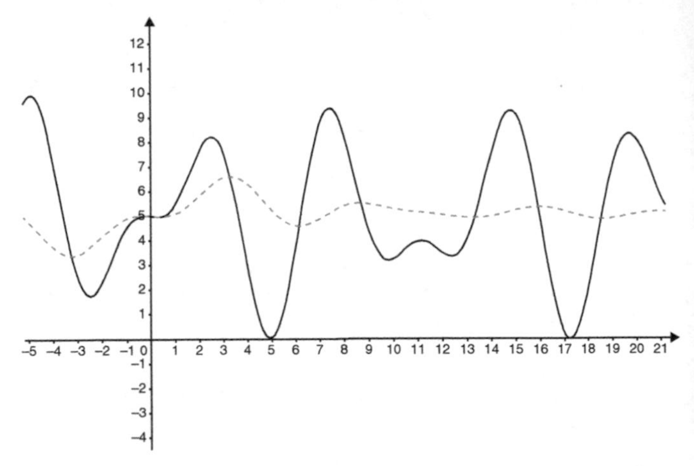

58. Tides for our fictional location. The dashed line represents the average tide from $t = 0$ to the current value of t.

We find A_0 from historical tide values (Figure 58) as follows. The two sine terms in the equation oscillate with an average value of zero, because each term spends the same amount of time below zero as it does above zero. So, to estimate A_0 we can simply compute the tide's average over time. The dotted curve in the figure represents the cumulative average tide values measured from $t = 0$; as we can see, in the long run the average value is settling down to $A_0 = 5$.

Now we know that our tide function is

$$f(t) = 5 + A_1 \sin(t) + A_2 \sin(1.5678t).$$

Our next task, finding A_1 and A_2, is accomplished in an extremely clever way. To find A_1, multiply the equation through by $\sin(t)$:

$$f(t) \cdot \sin(t) = 5\sin(t) + A_1 \sin^2(t) + A_2 \sin(1.5678t) \cdot \sin(t).$$

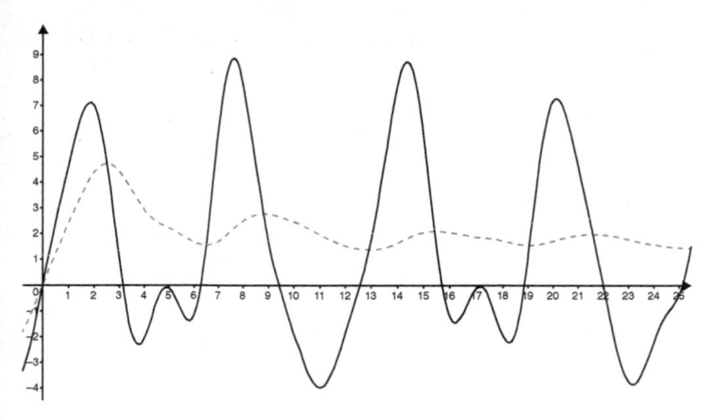

59. Finding the value of parameter A_1.

What is the long-term average of this new function? Well, $5\sin(t)$ is a sine wave spending just as much time above zero as it does below it, so its average is zero. The term $\sin(1.5678t)\cdot\sin(t)$ is a product of two sine waves. They both spend the same amount of time above and below zero, and since they cycle up and down with frequencies that are unrelated to each other, their product also averages to zero in the long run. Thus, the average value of $f(t)\cdot\sin(t)$ is equal to the average value of $A_1\sin^2(t)$. Pulling a trick from calculus, we know that the average value of $A_1\sin^2(t)$ is $A_1/2$. So, all we have to do to find A_1 is find the long-term average of $f(t)\cdot\sin(t)$, and multiply the result by 2. This is done in Figure 59. As we can see, the average is gradually decreasing. If we let the graph continue far enough to the right, we see that it levels off at 1.5. Therefore, $A_1 = 1.5\cdot 2 = 3$.

Now we have

$$f(t) = 5 + 3\sin(t) + A_2\sin(1.5678t),$$

and all that remains is to find A_2. We do this the same way we found A_1, this time multiplying $f(t)$ through by $\sin(1.5678t)$.

We get the result $A_2 = -2$, and we have found our function for the tide:

$$f(t) = 5 + 3\sin(t) - 2\sin(1.5678t).$$

Of course, the predictions generated by our formula will become less reliable as time goes on. But when the predictions start to waver, we can begin afresh and repeat the analysis.

In our simplified case, we broke the tide into two sine waves; in real practical situations, tide predictors use around twenty-five components. The amount of calculation required to generate a complete tide formula is staggering. Although today's computers handle it easily, in Thomson's day even a scientific calculator was still a pipe dream. Fortunately, necessity is the mother of invention. Thomson's brother James invented a machine, the *harmonic analyser* (Figure 60a), that computes the coefficients of the trigonometric series for the various frequencies produced by the interactions of the Earth, Sun, and Moon. Each of eleven 'ball and disc integrators' determines parameters for a particular frequency. Once the harmonic analyser did its job, the coefficients were passed to the *tide predicter* (Figure 60b). This device essentially did the job of a graphing calculator, plotting values of the tide function that the harmonic analyser had generated. Each pulley and crank corresponds to the sines and cosines for a certain frequency. Joined together as in the figure, the pulleys and cranks draw a graph that predicts the tides into the future as long as one wants. Thomson's invention and its successors were used to predict the tides in this way until at least the 1960s.

We conclude by noting a subtle but fundamental conceptual shift that has happened within this chapter. Until now we have been considering trigonometric functions to be *geometric* entities, that is, ratios of lengths within right triangles. When solving triangles, we considered the functions' geometric definitions. When we built

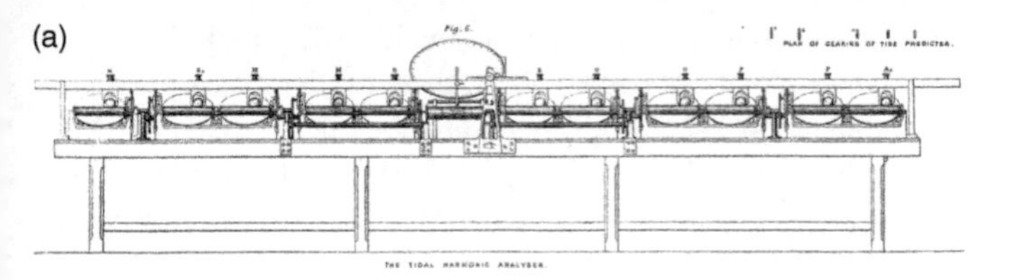

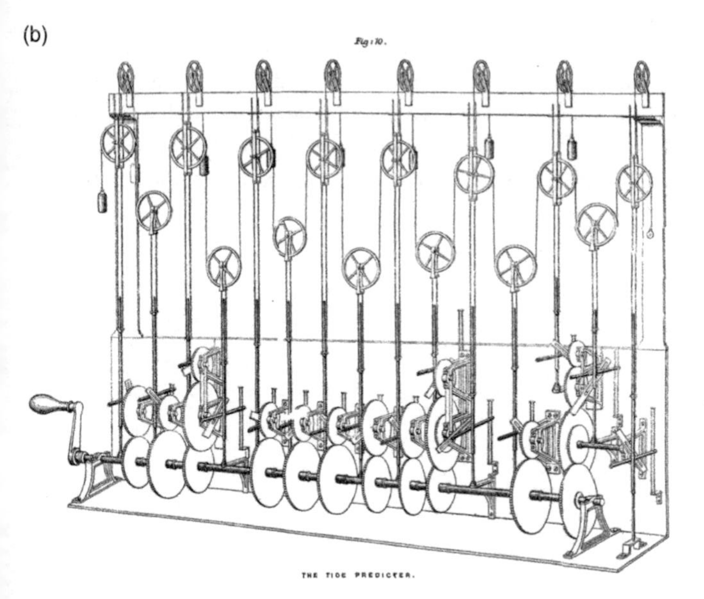

60. The Thomson brothers' harmonic analyser and tide predicter.

trigonometric tables, we used geometric theorems. When we proved identities sometimes we strayed into algebra, but the underlying truth was still geometric. In this chapter, we have almost forgotten that the sine and cosine are geometric entities at all. For the tides, each sine and cosine is simply a quantity that oscillates between its extreme values of −1 and 1. While our work has geometric implications, the mathematics itself is primarily *algebraic* or *analytic*.

This new perspective reflects a major shift within all of mathematics that occurred in the 17th and 18th centuries with the invention of Cartesian coordinates and the calculus. Geometry gradually changed from being *synthetic* (in the style of Euclid) to being *analytic*. The basic mathematical objects—originally points, lines, and shapes as well as numbers—became functions that accept input quantities and produce output quantities. This new approach, which is very powerful when applied to our physical world, has become the basis of mathematics as early as middle school. Although trigonometry was a latecomer to this change, it soon became a significant contributor. In Chapter 6 we see what happened to push trigonometry across the synthetic–analytic divide.

Chapter 6
... and beyond, to complex things

Brook Taylor's series are remarkable in many ways. One of the earliest of the Taylor series found in the 18th century is also one of the simplest:

TAYLOR SERIES FOR THE EXPONENTIAL FUNCTION:

$$e^x = 1 + \frac{x}{1!} + \frac{x^2}{2!} + \frac{x^3}{3!} + \frac{x^4}{4!} + \frac{x^5}{5!} + \dots$$

e is an irrational number equal to 2.718281828..., one of the most famous, yet enigmatic numbers in mathematics. Almost as important as π, it isn't nearly as easy to define. One way is to consider a bank account with an initial balance of $1, and an annual interest rate of 100 per cent. (If anyone finds such an account, be sure to let me know.) If the interest is calculated at the end of the year, we will have $2. If we compound monthly, the power of compound interest increases our total after a year to $2.61. If we compounded daily, we would have $2.71. The more frequently we compound, the more money we have—but it doesn't grow without bound, getting closer and closer to $2.718281828, or *e*. You can also define *e* as the sum

$$e = 1 + \frac{1}{1!} + \frac{1}{2!} + \frac{1}{3!} + \frac{1}{4!} + \frac{1}{5!} + \dots,$$

which is what you get when you substitute $x = 1$ into the Taylor series above. A persistent rumour asserts that the discoverer of e, Leonhard Euler, named it after himself. Don't believe it. The probable truth is more mundane: he introduced the letter e in his article 'Meditation on experiments made recently on the firing of cannon', likely because he hadn't yet used that letter to represent anything else in the article.

The Taylor series for e^x looks oddly similar to the series for the sine and cosine:

TAYLOR SERIES FOR THE SINE:

$$\sin\theta = \theta - \frac{\theta^3}{3!} + \frac{\theta^5}{5!} - \frac{\theta^7}{7!} + \frac{\theta^9}{9!} - \frac{\theta^{11}}{11!} + \frac{\theta^{13}}{13!} - \ldots$$

TAYLOR SERIES FOR THE COSINE:

$$\cos\theta = 1 - \frac{\theta^2}{2!} + \frac{\theta^4}{4!} - \frac{\theta^6}{6!} + \frac{\theta^8}{8!} - \frac{\theta^{10}}{10!} + \frac{\theta^{12}}{12!} - \ldots$$

If we add together the sine and cosine series, they fold together like a zipper and *almost* match the series for e^x :

$$\cos\theta + \sin\theta = 1 + \theta - \frac{\theta^2}{2!} - \frac{\theta^3}{3!} + \frac{\theta^4}{4!} + \frac{\theta^5}{5!} - \frac{\theta^6}{6!} - \frac{\theta^7}{7!} + \ldots$$

If it weren't for the alternating + and – signs, it would be a perfect match. That would have been very strange, because sines and cosines reside in the world of geometry, while exponential functions deal with the entirely different phenomena of growth and decay. Perhaps it's just a coincidence.

Early 18th-century mathematicians were not so sure.
As calculus was starting to become established, other curious parallels between the apparently disparate worlds of trigonometry and exponential functions were starting to appear. For instance, from calculus, consider the following pair of integrals:

$$\int \frac{1}{\sqrt{1-x^2}}\,dx = \sin^{-1} x$$

$$\int \frac{1}{\sqrt{x^2-1}}\,dx = \ln(\sqrt{x^2-1}+x).$$

The only difference between the left sides of these equations is $\sqrt{1-x^2}$ versus $\sqrt{x^2-1}$. We can use the first integral only when x is between –1 and 1, and the second integral only when x isn't between –1 and 1: otherwise, we would be taking the square root of a negative number. Nevertheless, the first of these almost identical integrals produces an inverse trigonometric function, while the second produces a natural logarithm—the inverse function of e^x. Roger Cotes (1682–1716) noticed this peculiarity when compiling lists of integrals in his *Harmonia mensuram*, so named because of this and other symmetries that he found.

Euler's formula and identity

If it were somehow possible to take the square root of a negative number, then, we would achieve the unlikeliest of mathematical marriages. Two centuries earlier, this unthinkable possibility had been considered already in the study of solutions to certain cubic equations. By allowing 'imaginary' quantities to exist, at least temporarily, the equations could be solved where ordinary algebra failed. But it had been just a game: before the solutions could be called 'real', one always had to return to the world of ordinary numbers. Allowing these imaginary numbers to exist, even temporarily, was a controversial and even dangerous move. Mathematics is not a subject where one can just assert that something exists when it doesn't.

But adventure never comes without a little risk, so let's fantasize and see what happens. Call i ('imaginary') the square root of –1. Clearly $i^2 = -1$. We can find higher powers of i easily enough: i^3 is equal to $i^2 \cdot i = -i$, then i^4 is equal to $i^2 \cdot i^2 = -1 \cdot -1 = 1$, and so on. Continuing in this manner, the powers of i turn out to be as follows:

$$i, -1, -i, 1, i, -1, -i, 1, i, -1, -i, 1, \ldots$$

The pattern of + and – terms in this sequence should look suspiciously familiar—it's what we saw when we combined the Taylor series for the sine and cosine. We can use this pattern in a very clever way: let $x = i\theta$ in the Taylor series for e^x.

$$\begin{aligned} e^{i\theta} &= 1 + i\theta + \frac{i^2\theta^2}{2!} + \frac{i^3\theta^3}{3!} + \frac{i^4\theta^4}{4!} + \frac{i^5\theta^5}{5!} + \frac{i^6\theta^6}{6!} + \frac{i^7\theta^7}{7!} + \ldots \\ &= 1 + i\theta - \frac{\theta^2}{2!} - i\frac{\theta^3}{3!} + \frac{\theta^4}{4!} + i\frac{\theta^5}{5!} - \frac{\theta^6}{6!} - i\frac{\theta^7}{7!} + \ldots \end{aligned}$$

If we separate the terms involving i from the others, a miracle occurs:

$$e^{i\theta} = \left(1 - \frac{\theta^2}{2!} + \frac{\theta^4}{4!} - \frac{\theta^6}{6!} + \ldots\right) + i\left(\theta - \frac{\theta^3}{3!} + \frac{\theta^5}{5!} - \frac{\theta^7}{7!} \ldots\right).$$

Our series has 'unzipped' perfectly into the Taylor series for cosine and sine! The nuptials of the trigonometric and exponential functions have now been accomplished in *Euler's formula*:

$$e^{i\theta} = \cos\theta + i\sin\theta.$$

Even better: if we substitute $\theta = \pi$, we arrive at the equation usually considered to be the most beautiful in all of mathematics: $e^{i\pi} = -1$, or *Euler's identity*:

$$e^{i\pi} + 1 = 0.$$

This formula contains the five most important numbers in mathematics (0, 1, e, π, and i) and the three most important operations (addition, multiplication, and exponentiation)—with the shocking benefit that the equation tying together all of these apparently unrelated concepts is true.

We can extend the fantasy even further. If we put $-i\theta$ into the exponent of Euler's formula, we get

$$e^{-i\theta} = \cos\theta - i\sin\theta.$$

Combine this with the original Euler's formula, do a little algebra, and we have

$$\cos\theta = \frac{e^{i\theta} + e^{-i\theta}}{2} \text{ and } \sin\theta = \frac{e^{i\theta} - e^{-i\theta}}{2i}.$$

In other words, we can now redefine the cosine and sine entirely, using exponentials—provided that we allow them to be raised to imaginary powers.

One might wonder, what is the good of all this game playing? Does it help us do trigonometry any better? In fact, it does. For instance, we can derive some identities more quickly this way. Consider the product-to-sum identities we saw in Chapter 4; back then we derived them by relying on the sine and cosine angle sum and difference laws. Now we can find them directly with a little algebra. For instance,

$$\begin{aligned}
\sin\alpha\sin\beta &= \frac{e^{i\alpha} - e^{-i\alpha}}{2i} \cdot \frac{e^{i\beta} - e^{-i\beta}}{2i} \\
&= \frac{e^{i(\alpha+\beta)} - e^{i(\alpha-\beta)} - e^{-i(\alpha-\beta)} + e^{-i(\alpha+\beta)}}{-4} \\
&= \frac{1}{2}\left(\frac{\left(e^{i(\alpha-\beta)} + e^{-i(\alpha-\beta)}\right) - \left(e^{i(\alpha+\beta)} + e^{-i(\alpha+\beta)}\right)}{2}\right) \\
&= \frac{1}{2}\left(\cos(\alpha-\beta) - \cos(\alpha+\beta)\right).
\end{aligned}$$

We will leave it as a challenge to derive the sine and cosine angle sum and difference laws from Euler's formula. They're even easier.

We are still left with a disturbing question: do these imaginary fantasies have any basis in reality? We seem to be building a house of cards: by simply assuming that *i* exists, which seems obviously impossible, we are basing our entire construction on a likely fallacy. It may not console you much to realize that we have been doing this sort of thing for millennia. For instance, what is the number $\sqrt{2}$? Obviously, it is 'the number whose square is 2'. But how do we know there is such a number? The common reply is to point to the right triangle whose two shorter sides have length 1; its hypotenuse has length $\sqrt{2}$. However, this only deepens the mystery. Given a unit length, there is a *line segment* that satisfies the condition, but that doesn't imply that there is a *number* that does. The ancient Greeks did not consider irrational magnitudes to be numbers at all, accepting only whole numbers and their ratios. $\sqrt{2}$ was a line segment, not a ratio. So we've been here before. In fact, speaking of mathematical fantasies, more recently we've gone much further than just imaginary numbers. To learn more, look up quaternions and octonions.

The Argand diagram and De Moivre's formula

Imaginary numbers are now at the heart of science and technology, used in the study of electromagnetic waves, cellular and wireless technologies, and fluid dynamics. It is thus my duty to make imaginary numbers a little more real, a little more trustworthy, than just saying 'we've placed our faith in impossible things before'. A picture can help; as they say, seeing is believing. The visual representation we use today was discovered first by Norwegian mathematician-cartographer Caspar Wessel (1745–1818); nine years later by the otherwise obscure Jean-Robert Argand (1768–1822); and finally by Abbé Adrien-Quentin Buée (1748–1826) in the same year. Clearly, the notion was in the air. All three discoverers almost suffered the fate of being ignored. Wessel is now considered to be the original discoverer, but the diagram is still named after Argand.

ANIMAL RIGHTS David DeGrazia
THE ANTARCTIC Klaus Dodds
ANTHROPOCENE Erle C. Ellis
ANTISEMITISM Steven Beller
ANXIETY Daniel Freeman and Jason Freeman
THE APOCRYPHAL GOSPELS Paul Foster
APPLIED MATHEMATICS Alain Goriely
ARCHAEOLOGY Paul Bahn
ARCHITECTURE Andrew Ballantyne
ARISTOCRACY William Doyle
ARISTOTLE Jonathan Barnes
ART HISTORY Dana Arnold
ART THEORY Cynthia Freeland
ARTIFICIAL INTELLIGENCE Margaret A. Boden
ASIAN AMERICAN HISTORY Madeline Y. Hsu
ASTROBIOLOGY David C. Catling
ASTROPHYSICS James Binney
ATHEISM Julian Baggini
THE ATMOSPHERE Paul I. Palmer
AUGUSTINE Henry Chadwick
AUSTRALIA Kenneth Morgan
AUTISM Uta Frith
AUTOBIOGRAPHY Laura Marcus
THE AVANT GARDE David Cottington
THE AZTECS Davíd Carrasco
BABYLONIA Trevor Bryce
BACTERIA Sebastian G. B. Amyes
BANKING John Goddard and John O. S. Wilson
BARTHES Jonathan Culler
THE BEATS David Sterritt
BEAUTY Roger Scruton
BEHAVIOURAL ECONOMICS Michelle Baddeley
BESTSELLERS John Sutherland
THE BIBLE John Riches
BIBLICAL ARCHAEOLOGY Eric H. Cline
BIG DATA Dawn E. Holmes
BIOGRAPHY Hermione Lee
BIOMETRICS Michael Fairhurst
BLACK HOLES Katherine Blundell
BLOOD Chris Cooper
THE BLUES Elijah Wald
THE BODY Chris Shilling
THE BOOK OF COMMON PRAYER Brian Cummings
THE BOOK OF MORMON Terryl Givens
BORDERS Alexander C. Diener and Joshua Hagen
THE BRAIN Michael O'Shea
BRANDING Robert Jones
THE BRICS Andrew F. Cooper
THE BRITISH CONSTITUTION Martin Loughlin
THE BRITISH EMPIRE Ashley Jackson
BRITISH POLITICS Anthony Wright
BUDDHA Michael Carrithers
BUDDHISM Damien Keown
BUDDHIST ETHICS Damien Keown
BYZANTIUM Peter Sarris
C. S. LEWIS James Como
CALVINISM Jon Balserak
CANCER Nicholas James
CAPITALISM James Fulcher
CATHOLICISM Gerald O'Collins
CAUSATION Stephen Mumford and Rani Lill Anjum
THE CELL Terence Allen and Graham Cowling
THE CELTS Barry Cunliffe
CHAOS Leonard Smith
CHARLES DICKENS Jenny Hartley
CHEMISTRY Peter Atkins
CHILD PSYCHOLOGY Usha Goswami
CHILDREN'S LITERATURE Kimberley Reynolds
CHINESE LITERATURE Sabina Knight
CHOICE THEORY Michael Allingham
CHRISTIAN ART Beth Williamson
CHRISTIAN ETHICS D. Stephen Long
CHRISTIANITY Linda Woodhead
CIRCADIAN RHYTHMS Russell Foster and Leon Kreitzman
CITIZENSHIP Richard Bellamy
CIVIL ENGINEERING David Muir Wood
CLASSICAL LITERATURE William Allan
CLASSICAL MYTHOLOGY Helen Morales
CLASSICS Mary Beard and John Henderson
CLAUSEWITZ Michael Howard
CLIMATE Mark Maslin
CLIMATE CHANGE Mark Maslin

CLINICAL PSYCHOLOGY Susan Llewelyn and Katie Aafjes-van Doorn
COGNITIVE NEUROSCIENCE Richard Passingham
THE COLD WAR Robert McMahon
COLONIAL AMERICA Alan Taylor
COLONIAL LATIN AMERICAN LITERATURE Rolena Adorno
COMBINATORICS Robin Wilson
COMEDY Matthew Bevis
COMMUNISM Leslie Holmes
COMPARATIVE LITERATURE Ben Hutchinson
COMPLEXITY John H. Holland
THE COMPUTER Darrel Ince
COMPUTER SCIENCE Subrata Dasgupta
CONCENTRATION CAMPS Dan Stone
CONFUCIANISM Daniel K. Gardner
THE CONQUISTADORS Matthew Restall and Felipe Fernández-Armesto
CONSCIENCE Paul Strohm
CONSCIOUSNESS Susan Blackmore
CONTEMPORARY ART Julian Stallabrass
CONTEMPORARY FICTION Robert Eaglestone
CONTINENTAL PHILOSOPHY Simon Critchley
COPERNICUS Owen Gingerich
CORAL REEFS Charles Sheppard
CORPORATE SOCIAL RESPONSIBILITY Jeremy Moon
CORRUPTION Leslie Holmes
COSMOLOGY Peter Coles
COUNTRY MUSIC Richard Carlin
CRIME FICTION Richard Bradford
CRIMINAL JUSTICE Julian V. Roberts
CRIMINOLOGY Tim Newburn
CRITICAL THEORY Stephen Eric Bronner
THE CRUSADES Christopher Tyerman
CRYPTOGRAPHY Fred Piper and Sean Murphy
CRYSTALLOGRAPHY A. M. Glazer
THE CULTURAL REVOLUTION Richard Curt Kraus
DADA AND SURREALISM David Hopkins
DANTE Peter Hainsworth and David Robey
DARWIN Jonathan Howard
THE DEAD SEA SCROLLS Timothy H. Lim
DECADENCE David Weir
DECOLONIZATION Dane Kennedy
DEMOCRACY Bernard Crick
DEMOGRAPHY Sarah Harper
DEPRESSION Jan Scott and Mary Jane Tacchi
DERRIDA Simon Glendinning
DESCARTES Tom Sorell
DESERTS Nick Middleton
DESIGN John Heskett
DEVELOPMENT Ian Goldin
DEVELOPMENTAL BIOLOGY Lewis Wolpert
THE DEVIL Darren Oldridge
DIASPORA Kevin Kenny
DICTIONARIES Lynda Mugglestone
DINOSAURS David Norman
DIPLOMACY Joseph M. Siracusa
DOCUMENTARY FILM Patricia Aufderheide
DREAMING J. Allan Hobson
DRUGS Les Iversen
DRUIDS Barry Cunliffe
DYNASTY Jeroen Duindam
DYSLEXIA Margaret J. Snowling
EARLY MUSIC Thomas Forrest Kelly
THE EARTH Martin Redfern
EARTH SYSTEM SCIENCE Tim Lenton
ECONOMICS Partha Dasgupta
EDUCATION Gary Thomas
EGYPTIAN MYTH Geraldine Pinch
EIGHTEENTH-CENTURY BRITAIN Paul Langford
THE ELEMENTS Philip Ball
EMOTION Dylan Evans
EMPIRE Stephen Howe
ENERGY SYSTEMS Nick Jenkins
ENGELS Terrell Carver
ENGINEERING David Blockley
THE ENGLISH LANGUAGE Simon Horobin
ENGLISH LITERATURE Jonathan Bate

THE ENLIGHTENMENT John Robertson
ENTREPRENEURSHIP Paul Westhead and Mike Wright
ENVIRONMENTAL ECONOMICS Stephen Smith
ENVIRONMENTAL ETHICS Robin Attfield
ENVIRONMENTAL LAW Elizabeth Fisher
ENVIRONMENTAL POLITICS Andrew Dobson
EPICUREANISM Catherine Wilson
EPIDEMIOLOGY Rodolfo Saracci
ETHICS Simon Blackburn
ETHNOMUSICOLOGY Timothy Rice
THE ETRUSCANS Christopher Smith
EUGENICS Philippa Levine
THE EUROPEAN UNION Simon Usherwood and John Pinder
EUROPEAN UNION LAW Anthony Arnull
EVOLUTION Brian and Deborah Charlesworth
EXISTENTIALISM Thomas Flynn
EXPLORATION Stewart A. Weaver
EXTINCTION Paul B. Wignall
THE EYE Michael Land
FAIRY TALE Marina Warner
FAMILY LAW Jonathan Herring
FASCISM Kevin Passmore
FASHION Rebecca Arnold
FEDERALISM Mark J. Rozell and Clyde Wilcox
FEMINISM Margaret Walters
FILM Michael Wood
FILM MUSIC Kathryn Kalinak
FILM NOIR James Naremore
THE FIRST WORLD WAR Michael Howard
FOLK MUSIC Mark Slobin
FOOD John Krebs
FORENSIC PSYCHOLOGY David Canter
FORENSIC SCIENCE Jim Fraser
FORESTS Jaboury Ghazoul
FOSSILS Keith Thomson
FOUCAULT Gary Gutting
THE FOUNDING FATHERS R. B. Bernstein
FRACTALS Kenneth Falconer
FREE SPEECH Nigel Warburton
FREE WILL Thomas Pink
FREEMASONRY Andreas Önnerfors
FRENCH LITERATURE John D. Lyons
THE FRENCH REVOLUTION William Doyle
FREUD Anthony Storr
FUNDAMENTALISM Malise Ruthven
FUNGI Nicholas P. Money
THE FUTURE Jennifer M. Gidley
GALAXIES John Gribbin
GALILEO Stillman Drake
GAME THEORY Ken Binmore
GANDHI Bhikhu Parekh
GARDEN HISTORY Gordon Campbell
GENES Jonathan Slack
GENIUS Andrew Robinson
GENOMICS John Archibald
GEOFFREY CHAUCER David Wallace
GEOGRAPHY John Matthews and David Herbert
GEOLOGY Jan Zalasiewicz
GEOPHYSICS William Lowrie
GEOPOLITICS Klaus Dodds
GERMAN LITERATURE Nicholas Boyle
GERMAN PHILOSOPHY Andrew Bowie
GLACIATION David J. A. Evans
GLOBAL CATASTROPHES Bill McGuire
GLOBAL ECONOMIC HISTORY Robert C. Allen
GLOBALIZATION Manfred Steger
GOD John Bowker
GOETHE Ritchie Robertson
THE GOTHIC Nick Groom
GOVERNANCE Mark Bevir
GRAVITY Timothy Clifton
THE GREAT DEPRESSION AND THE NEW DEAL Eric Rauchway
HABERMAS James Gordon Finlayson
THE HABSBURG EMPIRE Martyn Rady
HAPPINESS Daniel M. Haybron
THE HARLEM RENAISSANCE Cheryl A. Wall
THE HEBREW BIBLE AS LITERATURE Tod Linafelt
HEGEL Peter Singer
HEIDEGGER Michael Inwood

THE HELLENISTIC AGE Peter Thonemann
HEREDITY John Waller
HERMENEUTICS Jens Zimmermann
HERODOTUS Jennifer T. Roberts
HIEROGLYPHS Penelope Wilson
HINDUISM Kim Knott
HISTORY John H. Arnold
THE HISTORY OF ASTRONOMY Michael Hoskin
THE HISTORY OF CHEMISTRY William H. Brock
THE HISTORY OF CHILDHOOD James Marten
THE HISTORY OF CINEMA Geoffrey Nowell-Smith
THE HISTORY OF LIFE Michael Benton
THE HISTORY OF MATHEMATICS Jacqueline Stedall
THE HISTORY OF MEDICINE William Bynum
THE HISTORY OF PHYSICS J. L. Heilbron
THE HISTORY OF TIME Leofranc Holford-Strevens
HIV AND AIDS Alan Whiteside
HOBBES Richard Tuck
HOLLYWOOD Peter Decherney
THE HOLY ROMAN EMPIRE Joachim Whaley
HOME Michael Allen Fox
HOMER Barbara Graziosi
HORMONES Martin Luck
HUMAN ANATOMY Leslie Klenerman
HUMAN EVOLUTION Bernard Wood
HUMAN RIGHTS Andrew Clapham
HUMANISM Stephen Law
HUME A. J. Ayer
HUMOUR Noël Carroll
THE ICE AGE Jamie Woodward
IDENTITY Florian Coulmas
IDEOLOGY Michael Freeden
THE IMMUNE SYSTEM Paul Klenerman
INDIAN CINEMA Ashish Rajadhyaksha
INDIAN PHILOSOPHY Sue Hamilton
THE INDUSTRIAL REVOLUTION Robert C. Allen
INFECTIOUS DISEASE Marta L. Wayne and Benjamin M. Bolker
INFINITY Ian Stewart
INFORMATION Luciano Floridi
INNOVATION Mark Dodgson and David Gann
INTELLECTUAL PROPERTY Siva Vaidhyanathan
INTELLIGENCE Ian J. Deary
INTERNATIONAL LAW Vaughan Lowe
INTERNATIONAL MIGRATION Khalid Koser
INTERNATIONAL RELATIONS Paul Wilkinson
INTERNATIONAL SECURITY Christopher S. Browning
IRAN Ali M. Ansari
ISLAM Malise Ruthven
ISLAMIC HISTORY Adam Silverstein
ISOTOPES Rob Ellam
ITALIAN LITERATURE Peter Hainsworth and David Robey
JESUS Richard Bauckham
JEWISH HISTORY David N. Myers
JOURNALISM Ian Hargreaves
JUDAISM Norman Solomon
JUNG Anthony Stevens
KABBALAH Joseph Dan
KAFKA Ritchie Robertson
KANT Roger Scruton
KEYNES Robert Skidelsky
KIERKEGAARD Patrick Gardiner
KNOWLEDGE Jennifer Nagel
THE KORAN Michael Cook
KOREA Michael J. Seth
LAKES Warwick F. Vincent
LANDSCAPE ARCHITECTURE Ian H. Thompson
LANDSCAPES AND GEOMORPHOLOGY Andrew Goudie and Heather Viles
LANGUAGES Stephen R. Anderson
LATE ANTIQUITY Gillian Clark
LAW Raymond Wacks
THE LAWS OF THERMODYNAMICS Peter Atkins
LEADERSHIP Keith Grint
LEARNING Mark Haselgrove
LEIBNIZ Maria Rosa Antognazza

LEO TOLSTOY Liza Knapp
LIBERALISM Michael Freeden
LIGHT Ian Walmsley
LINCOLN Allen C. Guelzo
LINGUISTICS Peter Matthews
LITERARY THEORY Jonathan Culler
LOCKE John Dunn
LOGIC Graham Priest
LOVE Ronald de Sousa
MACHIAVELLI Quentin Skinner
MADNESS Andrew Scull
MAGIC Owen Davies
MAGNA CARTA Nicholas Vincent
MAGNETISM Stephen Blundell
MALTHUS Donald Winch
MAMMALS T. S. Kemp
MANAGEMENT John Hendry
MAO Delia Davin
MARINE BIOLOGY Philip V. Mladenov
THE MARQUIS DE SADE John Phillips
MARTIN LUTHER Scott H. Hendrix
MARTYRDOM Jolyon Mitchell
MARX Peter Singer
MATERIALS Christopher Hall
MATHEMATICAL FINANCE Mark H. A. Davis
MATHEMATICS Timothy Gowers
MATTER Geoff Cottrell
THE MEANING OF LIFE Terry Eagleton
MEASUREMENT David Hand
MEDICAL ETHICS Michael Dunn and Tony Hope
MEDICAL LAW Charles Foster
MEDIEVAL BRITAIN John Gillingham and Ralph A. Griffiths
MEDIEVAL LITERATURE Elaine Treharne
MEDIEVAL PHILOSOPHY John Marenbon
MEMORY Jonathan K. Foster
METAPHYSICS Stephen Mumford
METHODISM William J. Abraham
THE MEXICAN REVOLUTION Alan Knight
MICHAEL FARADAY Frank A. J. L. James
MICROBIOLOGY Nicholas P. Money
MICROECONOMICS Avinash Dixit
MICROSCOPY Terence Allen
THE MIDDLE AGES Miri Rubin
MILITARY JUSTICE Eugene R. Fidell
MILITARY STRATEGY Antulio J. Echevarria II
MINERALS David Vaughan
MIRACLES Yujin Nagasawa
MODERN ARCHITECTURE Adam Sharr
MODERN ART David Cottington
MODERN CHINA Rana Mitter
MODERN DRAMA Kirsten E. Shepherd-Barr
MODERN FRANCE Vanessa R. Schwartz
MODERN INDIA Craig Jeffrey
MODERN IRELAND Senia Pašeta
MODERN ITALY Anna Cento Bull
MODERN JAPAN Christopher Goto-Jones
MODERN LATIN AMERICAN LITERATURE Roberto González Echevarría
MODERN WAR Richard English
MODERNISM Christopher Butler
MOLECULAR BIOLOGY Aysha Divan and Janice A. Royds
MOLECULES Philip Ball
MONASTICISM Stephen J. Davis
THE MONGOLS Morris Rossabi
MOONS David A. Rothery
MORMONISM Richard Lyman Bushman
MOUNTAINS Martin F. Price
MUHAMMAD Jonathan A. C. Brown
MULTICULTURALISM Ali Rattansi
MULTILINGUALISM John C. Maher
MUSIC Nicholas Cook
MYTH Robert A. Segal
NAPOLEON David Bell
THE NAPOLEONIC WARS Mike Rapport
NATIONALISM Steven Grosby
NATIVE AMERICAN LITERATURE Sean Teuton
NAVIGATION Jim Bennett
NAZI GERMANY Jane Caplan
NELSON MANDELA Elleke Boehmer
NEOLIBERALISM Manfred Steger and Ravi Roy
NETWORKS Guido Caldarelli and Michele Catanzaro

THE NEW TESTAMENT Luke Timothy Johnson
THE NEW TESTAMENT AS LITERATURE Kyle Keefer
NEWTON Robert Iliffe
NIETZSCHE Michael Tanner
NINETEENTH-CENTURY BRITAIN Christopher Harvie and H. C. G. Matthew
THE NORMAN CONQUEST George Garnett
NORTH AMERICAN INDIANS Theda Perdue and Michael D. Green
NORTHERN IRELAND Marc Mulholland
NOTHING Frank Close
NUCLEAR PHYSICS Frank Close
NUCLEAR POWER Maxwell Irvine
NUCLEAR WEAPONS Joseph M. Siracusa
NUMBERS Peter M. Higgins
NUTRITION David A. Bender
OBJECTIVITY Stephen Gaukroger
OCEANS Dorrik Stow
THE OLD TESTAMENT Michael D. Coogan
THE ORCHESTRA D. Kern Holoman
ORGANIC CHEMISTRY Graham Patrick
ORGANIZATIONS Mary Jo Hatch
ORGANIZED CRIME Georgios A. Antonopoulos and Georgios Papanicolaou
ORTHODOX CHRISTIANITY A. Edward Siecienski
PAGANISM Owen Davies
PAIN Rob Boddice
THE PALESTINIAN-ISRAELI CONFLICT Martin Bunton
PANDEMICS Christian W. McMillen
PARTICLE PHYSICS Frank Close
PAUL E. P. Sanders
PEACE Oliver P. Richmond
PENTECOSTALISM William K. Kay
PERCEPTION Brian Rogers
THE PERIODIC TABLE Eric R. Scerri
PHILOSOPHY Edward Craig
PHILOSOPHY IN THE ISLAMIC WORLD Peter Adamson
PHILOSOPHY OF BIOLOGY Samir Okasha
PHILOSOPHY OF LAW Raymond Wacks
PHILOSOPHY OF SCIENCE Samir Okasha
PHILOSOPHY OF RELIGION Tim Bayne
PHOTOGRAPHY Steve Edwards
PHYSICAL CHEMISTRY Peter Atkins
PHYSICS Sidney Perkowitz
PILGRIMAGE Ian Reader
PLAGUE Paul Slack
PLANETS David A. Rothery
PLANTS Timothy Walker
PLATE TECTONICS Peter Molnar
PLATO Julia Annas
POETRY Bernard O'Donoghue
POLITICAL PHILOSOPHY David Miller
POLITICS Kenneth Minogue
POPULISM Cas Mudde and Cristóbal Rovira Kaltwasser
POSTCOLONIALISM Robert Young
POSTMODERNISM Christopher Butler
POSTSTRUCTURALISM Catherine Belsey
POVERTY Philip N. Jefferson
PREHISTORY Chris Gosden
PRESOCRATIC PHILOSOPHY Catherine Osborne
PRIVACY Raymond Wacks
PROBABILITY John Haigh
PROGRESSIVISM Walter Nugent
PROJECTS Andrew Davies
PROTESTANTISM Mark A. Noll
PSYCHIATRY Tom Burns
PSYCHOANALYSIS Daniel Pick
PSYCHOLOGY Gillian Butler and Freda McManus
PSYCHOLOGY OF MUSIC Elizabeth Hellmuth Margulis
PSYCHOPATHY Essi Viding
PSYCHOTHERAPY Tom Burns and Eva Burns-Lundgren
PUBLIC ADMINISTRATION Stella Z. Theodoulou and Ravi K. Roy
PUBLIC HEALTH Virginia Berridge
PURITANISM Francis J. Bremer
THE QUAKERS Pink Dandelion

QUANTUM THEORY John Polkinghorne
RACISM Ali Rattansi
RADIOACTIVITY Claudio Tuniz
RASTAFARI Ennis B. Edmonds
READING Belinda Jack
THE REAGAN REVOLUTION Gil Troy
REALITY Jan Westerhoff
THE REFORMATION Peter Marshall
RELATIVITY Russell Stannard
RELIGION IN AMERICA Timothy Beal
THE RENAISSANCE Jerry Brotton
RENAISSANCE ART Geraldine A. Johnson
REPTILES T. S. Kemp
REVOLUTIONS Jack A. Goldstone
RHETORIC Richard Toye
RISK Baruch Fischhoff and John Kadvany
RITUAL Barry Stephenson
RIVERS Nick Middleton
ROBOTICS Alan Winfield
ROCKS Jan Zalasiewicz
ROMAN BRITAIN Peter Salway
THE ROMAN EMPIRE Christopher Kelly
THE ROMAN REPUBLIC David M. Gwynn
ROMANTICISM Michael Ferber
ROUSSEAU Robert Wokler
RUSSELL A. C. Grayling
RUSSIAN HISTORY Geoffrey Hosking
RUSSIAN LITERATURE Catriona Kelly
THE RUSSIAN REVOLUTION S. A. Smith
THE SAINTS Simon Yarrow
SAVANNAS Peter A. Furley
SCEPTICISM Duncan Pritchard
SCHIZOPHRENIA Chris Frith and Eve Johnstone
SCHOPENHAUER Christopher Janaway
SCIENCE AND RELIGION Thomas Dixon
SCIENCE FICTION David Seed
THE SCIENTIFIC REVOLUTION Lawrence M. Principe
SCOTLAND Rab Houston
SECULARISM Andrew Copson
SEXUAL SELECTION Marlene Zuk and Leigh W. Simmons
SEXUALITY Véronique Mottier
SHAKESPEARE'S COMEDIES Bart van Es
SHAKESPEARE'S SONNETS AND POEMS Jonathan F. S. Post
SHAKESPEARE'S TRAGEDIES Stanley Wells
SIKHISM Eleanor Nesbitt
THE SILK ROAD James A. Millward
SLANG Jonathon Green
SLEEP Steven W. Lockley and Russell G. Foster
SOCIAL AND CULTURAL ANTHROPOLOGY John Monaghan and Peter Just
SOCIAL PSYCHOLOGY Richard J. Crisp
SOCIAL WORK Sally Holland and Jonathan Scourfield
SOCIALISM Michael Newman
SOCIOLINGUISTICS John Edwards
SOCIOLOGY Steve Bruce
SOCRATES C. C. W. Taylor
SOUND Mike Goldsmith
SOUTHEAST ASIA James R. Rush
THE SOVIET UNION Stephen Lovell
THE SPANISH CIVIL WAR Helen Graham
SPANISH LITERATURE Jo Labanyi
SPINOZA Roger Scruton
SPIRITUALITY Philip Sheldrake
SPORT Mike Cronin
STARS Andrew King
STATISTICS David J. Hand
STEM CELLS Jonathan Slack
STOICISM Brad Inwood
STRUCTURAL ENGINEERING David Blockley
STUART BRITAIN John Morrill
SUPERCONDUCTIVITY Stephen Blundell
SUPERSTITION Stuart Vyse
SYMMETRY Ian Stewart
SYNAESTHESIA Julia Simner
SYNTHETIC BIOLOGY Jamie A. Davies
TAXATION Stephen Smith
TEETH Peter S. Ungar
TELESCOPES Geoff Cottrell
TERRORISM Charles Townshend
THEATRE Marvin Carlson
THEOLOGY David F. Ford

THINKING AND REASONING
Jonathan St B. T. Evans
THOMAS AQUINAS Fergus Kerr
THOUGHT Tim Bayne
TIBETAN BUDDHISM
Matthew T. Kapstein
TIDES David George Bowers and
Emyr Martyn Roberts
TOCQUEVILLE Harvey C. Mansfield
TOPOLOGY Richard Earl
TRAGEDY Adrian Poole
TRANSLATION Matthew Reynolds
THE TREATY OF VERSAILLES
Michael S. Neiberg
TRIGONOMETRY Glen Van Brummelen
THE TROJAN WAR Eric H. Cline
TRUST Katherine Hawley
THE TUDORS John Guy
TWENTIETH-CENTURY
BRITAIN Kenneth O. Morgan
TYPOGRAPHY Paul Luna
THE UNITED NATIONS
Jussi M. Hanhimäki
UNIVERSITIES AND COLLEGES
David Palfreyman and Paul Temple
THE U.S. CONGRESS Donald A. Ritchie
THE U.S. CONSTITUTION
David J. Bodenhamer
THE U.S. SUPREME COURT
Linda Greenhouse
UTILITARIANISM
Katarzyna de Lazari-Radek and
Peter Singer
UTOPIANISM Lyman Tower Sargent
VETERINARY SCIENCE James Yeates
THE VIKINGS Julian D. Richards
VIRUSES Dorothy H. Crawford
VOLTAIRE Nicholas Cronk
WAR AND TECHNOLOGY
Alex Roland
WATER John Finney
WAVES Mike Goldsmith
WEATHER Storm Dunlop
THE WELFARE STATE David Garland
WILLIAM SHAKESPEARE
Stanley Wells
WITCHCRAFT Malcolm Gaskill
WITTGENSTEIN A. C. Grayling
WORK Stephen Fineman
WORLD MUSIC Philip Bohlman
THE WORLD TRADE
ORGANIZATION Amrita Narlikar
WORLD WAR II Gerhard L. Weinberg
WRITING AND SCRIPT
Andrew Robinson
ZIONISM Michael Stanislawski

Available soon:

NIELS BOHR J. L. Heilbron
THE SUN Philip Judge
AERIAL WARFARE Frank Ledwidge
RENEWABLE ENERGY Nick Jelley
NUMBER THEORY
Robin Wilson

For more information visit our website

www.oup.com/vsi/

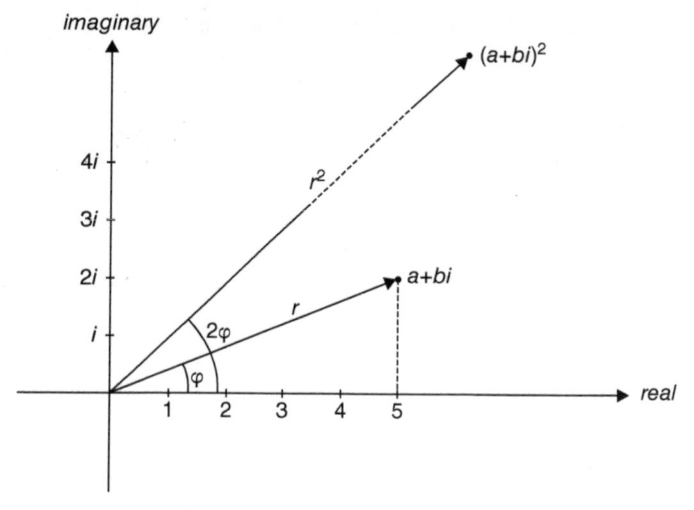

61. The Argand diagram. In this example, $a = 5$ and $b = 2$.

In Figure 61 the real numbers are plotted on the horizontal axis. We draw another axis perpendicular to it and place $i = \sqrt{-1}$ one unit above the origin. All the multiples of i, the imaginary numbers, appear on this vertical axis. With these axes we can locate on the page the sum of any real number and imaginary number, for instance $5 + 2i$ (indicated in the figure). Such combinations of real and imaginary numbers are called *complex numbers*. The real part (in our case, 5) is usually called a, and the imaginary part (in our case, 2) is usually called b.

This diagram allows us to locate the complex numbers on a plane, but it does much more. Think of each complex number as a vector (or arrow) emerging from the origin. The length r of this vector is called the *modulus*; in our case $r = \sqrt{5^2 + 2^2} = \sqrt{29} \approx 5.385$. The angle that the vector makes with the real number line, the *argument*, is $\varphi = \tan^{-1}\dfrac{b}{a}$; in our case $\varphi \approx 0.3805$ radians $= 21.8°$.

This gives us a new, compact way of writing complex numbers. Since $a = r\cos\varphi$ and $b = r\sin\varphi$, we can use Euler's formula:

$$a + bi = r\cos\varphi + i \cdot r\sin\varphi = re^{i\varphi}.$$

In our case, $5 + 2i \approx 5.385e^{0.3805i}$.

Our new notation explains why the Argand diagram is so useful. If we square our complex number $5 + 2i$, we get

$$(5 + 2i)^2 = 25 + 10i + 10i - 4 = 21 + 20i.$$

What does this squared quantity have to do with the original? Our new compact form allows us to see more clearly:

$$\left(5.385e^{0.3805i}\right)^2 = 5.385^2 \cdot e^{2 \cdot 0.3805i} = 29e^{0.7610i}.$$

So, the square of a complex number is a vector whose length (modulus) is the square of the original, but whose angle (argument) is *double* the original!

This arithmetic allows us to see that something similar will happen for the cube of a complex number, or a fourth power, or any other power. The modulus grows according to the cube, or fourth power, and so on; but the argument is tripled, quadrupled, and so on. This surprising property gives the Argand plane a good reason to be worthy of our attention; it gives us our first glimpse into the geometry of complex numbers.

The modulus and argument take us even further. Summarizing the above arithmetic for an arbitrary power n, we have

$$(r\cos\theta + i \cdot r\sin\theta)^n = \left(re^{i\theta}\right)^n = r^n e^{in\theta} = r^n(\cos n\theta + i\sin n\theta).$$

If we cancel r^n out of the left and right sides of this equation, we get *De Moivre's formula*:

$$(\cos\theta + i\sin\theta)^n = \cos n\theta + i\sin n\theta.$$

This seemingly innocuous equation gives us the capacity to turn some more geometry into algebra. Near the end of Chapter 3, we developed double-, triple-, and higher-angle formulas for the sine and cosine, for the purpose of constructing trigonometric tables. De Moivre's formula gives us a simple tool to generate these formulas in an entirely different way. For instance, for the triple-angle formulas, plug $n = 3$ into the left side of De Moivre's formula and expand:

$$(\cos\theta + i\sin\theta)^3 = \ldots = \cos^3\theta - 3\sin^2\theta\ \cos\theta$$
$$+ i\left(3\cos^2\theta\ \sin\theta - \sin^3\theta\right).$$

De Moivre tells us that the real part of this expression must be $\cos 3\theta$, and the imaginary part (the part multiplied by i) must be $\sin 3\theta$. Therefore

$$\cos 3\theta = \cos^3\theta - 3\sin^2\theta\cos\theta \quad \text{and} \quad \sin 3\theta = 3\cos^2\theta\sin\theta - \sin^3\theta.$$

If we replace $\sin^2\theta$ with $1-\cos^2\theta$ in the first expression and $\cos^2\theta$ with $1-\sin^2\theta$ in the second, we get both triple-angle formulas immediately:

$$\cos 3\theta = 4\cos^3\theta - 3\cos\theta \quad \text{and} \quad \sin 3\theta = 3\sin\theta - 4\sin^3\theta.$$

This works for any power n, as long as we have the patience (or the software) to do the algebra.

The hyperbolic functions

Let's return to our definitions of the cosine and sine in terms of exponentials and imaginary numbers:

$$\cos\theta = \frac{e^{i\theta} + e^{-i\theta}}{2} \quad \text{and} \quad \sin\theta = \frac{e^{i\theta} - e^{-i\theta}}{2i}.$$

Emboldened by our success in bringing imaginary numbers into the world, let's consider what happens if we allow the *angles* in these functions to be imaginary as well. The result is surprising: when the angles are imaginary, the imaginary numbers in the powers of the exponents disappear altogether!

$$\cos i\theta = \frac{e^{i(i\theta)} + e^{-i(i\theta)}}{2} = \frac{e^{-\theta} + e^{\theta}}{2}$$
$$\sin i\theta = \frac{e^{i(i\theta)} - e^{-i(i\theta)}}{2i} = \frac{e^{-\theta} - e^{\theta}}{2i}$$

A minor rearrangement of these equations gives us the standard definitions of the *hyperbolic trigonometric functions*:

$$\cosh\theta = \cos i\theta = \frac{e^{\theta} + e^{-\theta}}{2}$$
$$\sinh\theta = -i\sin i\theta = \frac{e^{\theta} - e^{-\theta}}{2}.$$

There are a couple of mysteries in these definitions: why bless these particular quantities with the names 'cosine' and 'sine'? And why call them 'hyperbolic'? These questions go back to the 1750s and 1760s. It will come as no surprise by now that once again, different people working independently of each other were working simultaneously on these problems. In this case they were Vincenzo Riccati (1707–75), son of Jacopo for whom the Riccati differential equation is named, and Johann Heinrich Lambert (1728–77), with some inspiration from a preliminary study by François Daviet de Foncenex (1734–99). Lambert was a colleague and friend of Euler late in their lives; he is best known for devising seven new map projections and for proving that π is irrational. Although Riccati was the first to publish on the hyperbolic functions, Lambert placed the subject in a mathematical context that attracted more attention from colleagues; it is his name that is most closely associated with the topic today.

It was in the publication proving π's irrationality, 'Mémoire sur quelques propriétés remarquables des quantités transcendantes circulaires et logarithmiques', that Lambert introduced the hyperbolic functions. It is here that we see why the terms 'sine', 'cosine', and 'hyperbolic' are used. Recall that the modern sine and cosine functions get their meanings from the unit circle $x^2 + y^2 = 1$ in the centre of Figure 62: in this circle, the cosine of θ is the x coordinate, and the sine of θ is the y coordinate. We know that $\cos^2\theta + \sin^2\theta = 1$. Does something similar happen when we square $\cosh a$ and $\sinh a$? Using our exponential definitions, we get

$$\cosh^2 a = \frac{1}{4}\left(e^{2a} + 2 + e^{-2a}\right) \quad \text{and} \quad \sinh^2 a = \frac{1}{4}\left(e^{2a} - 2 + e^{-2a}\right).$$

If we *subtract* one from the other, we end up with

$$\cosh^2 a - \sinh^2 a = 1,$$

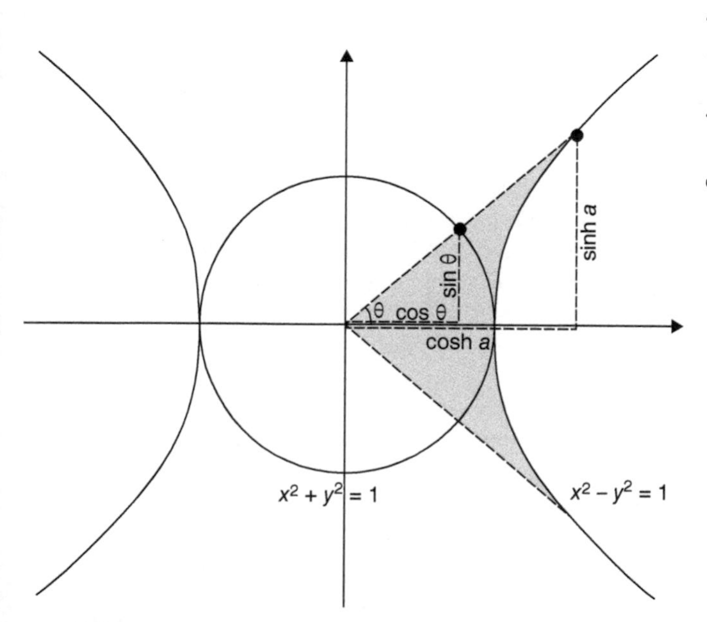

62. Defining the hyperbolic functions using the unit hyperbola.

almost the Pythagorean theorem, but with a sign switched. The cosh and sinh, then, lie not on the unit circle $x^2 + y^2 = 1$, but on the unit *hyperbola* $x^2 - y^2 = 1$ (Figure 62).

You might have noticed that I slipped in a change in the name of the argument of the hyperbolic trigonometric functions, from θ to a. We started our hyperbolic journey by considering what would happen if we evaluated $\cos(i\theta)$, which we ended up calling $\cosh\theta$. Originally θ was the angle from the x axis in the unit circle in Figure 62, but what does it become if we multiply by i? There's no reason why it still should be an angle at all. It turns out that the argument of the hyperbolic functions is no longer an angle, but rather a quantity known as the *magnitude of the hyperbolic angle*, the shaded region in Figure 62. Although the hyperbolic angle a and the ordinary angle θ no longer represent quite the same things, there is still a parallel between them. The area swept out within the unit circle when the angle increases from 0 to θ is equal to $\theta / 2$; the area swept out within the unit hyperbola when the hyperbolic angle increases from 0 to a is equal to $a / 2$.

There is a nice analogy, then, between the trigonometric and the hyperbolic functions. Is it enough to justify calling the new functions 'trigonometric'? Lambert thought so. In a later paper, he noticed that the corresponding tangent functions are also related. We can see right away from the two similar dashed triangles in Figure 62 that

$$\frac{\sin\theta}{\cos\theta} = \frac{\sinh a}{\cosh a}.$$

In other words, if we define 'tanh' in the usual way as 'sinh' divided by 'cosh', then $\tan\theta = \tanh a$. From this and other analogies between the trigonometric and hyperbolic worlds, Lambert asserted that there was 'nothing repugnant to the original meaning' of the trigonometric terms in his use of the terms 'hyperbolic sine' and 'hyperbolic cosine'.

The parallels continue when we consider the hyperbolic identities. For instance, consider the hyperbolic cosine sum law, which can be derived from the definition of the hyperbolic cosine:

$$\begin{aligned}\cosh(\alpha+\beta) &= \frac{e^{\alpha+\beta}+e^{-\alpha-\beta}}{2} = \frac{2e^{\alpha+\beta}+2e^{-\alpha-\beta}}{4} \\ &= \frac{\left(e^{\alpha+\beta}+e^{-\alpha-\beta}+e^{\beta-\alpha}+e^{\alpha-\beta}\right)+\left(e^{\alpha+\beta}+e^{-\alpha-\beta}-e^{\beta-\alpha}-e^{\alpha-\beta}\right)}{4} \\ &= \left(\frac{e^{\alpha}+e^{-\alpha}}{2}\right)\left(\frac{e^{\beta}+e^{-\beta}}{2}\right)+\left(\frac{e^{\alpha}-e^{-\alpha}}{2}\right)\left(\frac{e^{\beta}-e^{-\beta}}{2}\right) \\ &= \cosh\alpha\cosh\beta+\sinh\alpha\sinh\beta.\end{aligned}$$

Compare this with the ordinary cosine angle sum law,

$$\cos(\alpha+\beta)=\cos\alpha\cos\beta-\sin\alpha\sin\beta.$$

Something similar happens with the hyperbolic sine sum law:

$$\sinh(\alpha+\beta)=\cosh\alpha\sinh\beta+\sinh\alpha\cosh\beta;$$

compare with

$$\sin(\alpha+\beta)=\cos\alpha\sin\beta+\sin\alpha\cos\beta.$$

The list goes on and on. Here are a few other hyperbolic identities, listed with their trigonometric counterparts:

HYPERBOLIC SINE DOUBLE-ARGUMENT FORMULA:

$$\sinh 2x = 2\sinh x\cosh x$$
$$\sin 2x = 2\sin x\cos x$$

HYPERBOLIC SINE TRIPLE-ARGUMENT FORMULA:

$$\sinh 3x = 3\sinh x + 4\sinh^3 x$$
$$\sin 3x = 3\sin x - 4\sin^3 x$$

HYPERBOLIC SINE PRODUCT-TO-SUM FORMULA:

$$\sinh x \sinh y = \frac{1}{2}(\cosh(x+y) - \cosh(x-y))$$

$$\sin x \sin y = \frac{1}{2}(\cos(x-y) - \cos(x+y))$$

HYPERBOLIC SINE SUM-TO-PRODUCT FORMULA:

$$\sinh x + \sinh y = 2\sinh\left(\frac{x+y}{2}\right)\cosh\left(\frac{x-y}{2}\right)$$

$$\sin x + \sin y = 2\sin\frac{x+y}{2}\cos\frac{x-y}{2}$$

Each pair of formulas is essentially identical, except for an occasional sign change from + to – or vice versa. These sign changes were noticed by G. Osborn in a 1902 issue of the *Mathematical Gazette*, and have since been characterized as follows:

OSBORN'S RULE: Any trigonometric identity can be converted to a hyperbolic identity by converting sin/cos to sinh/cosh. Whenever a term contains a square of a sine, the sign in front of it should be negated.

For instance, consider the tangent sum formula:

$$\tan(\alpha+\beta) = \frac{\tan\alpha + \tan\beta}{1 - \tan\alpha\tan\beta}.$$

The two tangents in the numerator are both sines over cosines, so there are no *products* of sines. But the $\tan\alpha\tan\beta$ is $(\sin\alpha/\cos\alpha)\cdot(\sin\beta/\cos\beta)$. It contains a product of two sines, so Osborn tells us that the sign in front must be reversed. We get the correct formula,

$$\tanh(\alpha+\beta) = \frac{\tanh\alpha + \tanh\beta}{1 + \tanh\alpha\tanh\beta}.$$

The eerie parallels continue when we enter the world of calculus. Here are a few examples:

$$(\sin\theta)' = \cos\theta \qquad (\sinh a)' = \cosh a$$

$$(\cos\theta)' = -\sin\theta \qquad (\cosh a)' = \sinh a$$

$$(\tan\theta)' = \sec^2\theta \qquad (\tanh a)' = \operatorname{sech}^2 a$$

$$\int \tan x\, dx = -\ln|\cos x| + C \qquad \int \tanh a\, dx = \ln|\cosh a| + C$$

$$(\sin^{-1} x)' = \frac{1}{\sqrt{1-x^2}} \qquad (\sinh^{-1} x)' = \frac{1}{\sqrt{1+x^2}}$$

$$(\tan^{-1} x)' = \frac{1}{1+x^2} \qquad (\tanh^{-1} x)' = \frac{1}{1-x^2}$$

$$\int \sin^{-1} x\, dx = x\sin^{-1} x + \sqrt{1-x^2} + C \qquad \int \sinh^{-1} x\, dx = x\sinh^{-1} x - \sqrt{1+x^2} + C$$

If you weren't convinced before, perhaps you are now. The hyperbolic sines and cosines fully deserve their auspicious names.

Hanging out with the catenary

One of my friends, a welder and contractor, is very patient with me when I describe in awe the wonders of pure mathematics. In the end, though, he always brings me back to earth: what is all of this good for? Where do we see it in the world around us? It's an important question. The world is full of wonders, but someone has to build it. We would need some calculus to answer the question fully with respect to the hyperbolic functions, so we won't go into detail here. We begin with the observation that the exponential and trigonometric functions have simple relationships with their derivatives. For instance, e^x is its own derivative; and the second derivative of $\sin\theta$ is just $-\sin\theta$. Many real-life situations involve simple statements of rates of change: for instance, Newton's Law of Cooling asserts that the temperature of a cup of coffee sitting in a room cools down at a rate proportional to the difference between the coffee's temperature and the room temperature. It turns out that the simplicity of the derivative of e^x allows us to use it with Newton's Law of Cooling. In the end, we find that the coffee's temperature cools according to the laws of exponential decay.

Something similar works with the trigonometric functions. Consider a spring suspended from the ceiling with a weight attached. Set the spring in motion. The spring's movement is affected by three forces: gravity, the force pulling the spring back to equilibrium (Hooke's Law), and the force exerted by the air slowing down the spring (viscous damping). These forces combined are equal to mass times acceleration. Acceleration is the second derivative of position, and the second derivative of the sine is just the negative of the sine. So, just as with Newton's Law of Cooling, it's natural that the sine function enters into the spring's path; and it is the sine function that represents its vibration.

Next, consider a chain hanging between two poles, as in Figure 63. Three forces act on any given section of the cable: its own weight pulling downward, the tension force at the low point exerted horizontally, and the tension force at the upper point pulling in

63. A hanging chain.

the direction of the curve. When the chain is at rest, these three forces are in perfect balance with each other. Once more, this balance results in an equation representing the rate of change (the derivative) of the curve. Deriving the curve's equation goes beyond the scope of this book (see Further Reading). But once again, the simplicity of the derivatives of (in this case) the hyperbolic functions provides the key to the problem. The solution turns out to be $y = \cosh x$ (see Figure 64).

This curve, known as the *catenary* after the Latin word for 'chain', was studied throughout the 17th century. Galileo approximated the chain as a parabola; it wasn't until calculus became available that a variety of scientists and mathematicians (including Hooke, Leibniz, Huygens, and Johann Bernoulli) discovered the chain's true nature. None of these scholars used the language of the hyperbolic cosine; that was still almost a century in the future. Their solutions involved only the exponential representation of the function.

We turn our attention from the hanging chain to an apparently unrelated problem: that of constructing the optimal shape for an arch. If supported only by its own weight, an arch will be especially stable if it transforms the downward force of gravity into a compression force that presses in the direction of the arch's curve. If this were not the case, then gravity would be working

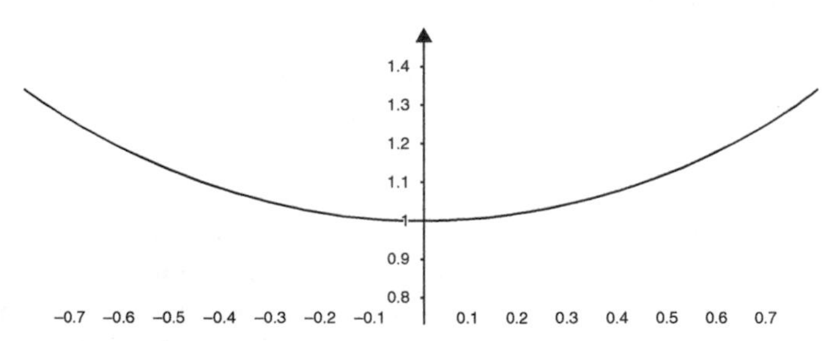

64. The graph of the hyperbolic cosine function, also called the catenary.

65. The Gateway Arch, St Louis, MO.

constantly to cause the arch to collapse. Robert Hooke (1635–1703) appears to have been the first to recognize the solution: the most stable arch is simply a catenary turned upside down. The force of gravity acting on the arch corresponds to the tension force acting on the hanging chain; in both cases, these forces are most stable when they point in the direction of the curve. Hooke published his work encrypted as a Latin anagram in 1675. In an age where 'publish or perish' was not the way of all academia, this was how Hooke could be the only person in the know, but also could claim priority if someone else were to discover the solution later. Two years after his death, his executor published the solution for all to see.

Catenary arches are found in a number of buildings around the world. The most famous is the Gateway Arch in St Louis (Figure 65). Others include Brunelleschi's dome in Florence, Italy; Budapest's train station; and the roof of Dulles Airport in Washington, DC. Catenaries have been discovered without the help of calculus by several cultures: they are found in Inuit igloos; mud huts in Cameroon; and the Tāq Kasrā, a 3rd–6th-century AD Sasanian vaulted hall in Iraq 37 metres high (Figure 66). The Tāq Kasrā was so stable that it is the only structure that remains

66. The Tāq Kasrā.

standing today in the ruins of the ancient city Ctesiphon. It is a fitting tribute to the universality and diversity of human knowledge that this chapter's exploration began with the discovery of imaginary numbers in early modern Europe, and ends with one of its most impressive realizations in a palace built in Iraq almost two millennia ago.

Chapter 7
Spheres and more

We left our ancient astronomer Hipparchus of Rhodes behind some while ago in Chapter 3, determining the eccentricity of the Sun's orbit around the Earth and computing sine tables geometrically. Our departure was well timed; Hipparchus' next step would have been out of our reach at that moment. Now, we're ready. If you are near a window or outdoors, look up. If we ignore the filters that education has been busy installing in our brains since we were children, it should be obvious to us that the Earth is at the centre of a very large dome: the sky. Now look down. The ground may appear to be flat, especially if we live on a prairie or on a boat at sea; but it has been known since ancient times that the Earth is also a sphere. (Don't believe the story about Christopher Columbus trying to convince the king of Spain that the Earth is round; it is a myth.) If you are on a coast, you can confirm this by watching a distant ship. If it is far enough away, the ship's hull appears to have sunk below the water. In fact, the hull has disappeared from sight below the curvature of the Earth's surface.

So, we live on a large sphere at the centre of an unfathomably larger sphere that contains the Sun, Moon, stars, and planets on its surface. Models of this *celestial sphere*, called *armillary spheres*, have been constructed for millennia; nowadays they tend to be found in botanical gardens or atop historical buildings (Figure 67). The celestial sphere is the arena within which the

67. An armillary sphere.

stars move, including Hipparchus' Sun. However, the trigonometry people use today—including everything so far in this book—exists on a very different arena: a flat surface such as a piece of paper or a blackboard. We are joining Hipparchus in virgin territory, and we will have to start over.

When we turn to the sphere, we must be prepared to give up much of what we thought we knew. If we walk on our sphere in a straight line for a long enough time, eventually we will trace out a *great circle*, returning to where we started. Unlike our familiar flat surface, we cannot travel as far as we like; the furthest we can get from our starting point is half a great circle away. Life on the sphere gets stranger still: imagine you and your friend walk away from the North Pole at right angles to each other. When you both reach the equator, you turn towards each other. When you meet, you will have traced out a triangle with three right angles (Figure 68). The sum of the angles of this triangle is not the usual 180°, but 270°.

Let's consider a couple of other triangles. Imagine we are standing on a Earth-sized sphere and we draw a small triangle on the ground. It is so small that it is indistinguishable from a plane triangle. The angle sum is greater than 180° but only by an immeasurably tiny amount, due to the very slight bulge of the spherical surface above the plane triangle joining the three vertices. Now consider three points on the equator, each separated by 120°. Join them, and we have a triangle—or is it a circle? In fact, it is both. The three 'angles' are all 180°, so the sum of the angles is 540°. The angle sums of spherical triangles can fall anywhere between the two extremes of 180° and 540°.

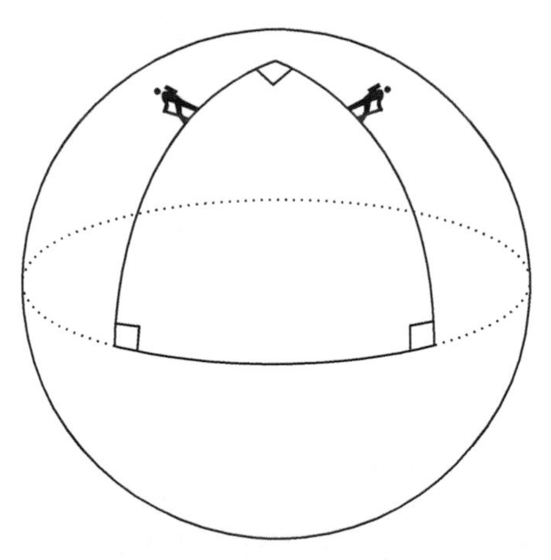

68. A 270° spherical triangle.

Measuring lengths can be even more disorienting. What are the side lengths of the triangle in Figure 68? There are two ways to answer this question. First, notice that each side is ¼ of a great circle, so its length is $\frac{1}{4}(2\pi r) = \pi r / 2$. Therefore, if we're on a unit sphere, the length is $\pi / 2$. Second, we can measure the side as the length of the arc, in this case 90°. Measuring side lengths in degrees is a marked change of perspective; it opens the door to a whole new trigonometric world.

We enter that door through the sky. Look again at the armillary sphere in Figure 67. You'll find several parallel circles on it; the largest one in the middle is called the *celestial equator*. In the northern hemisphere's night sky, the equator rises from the east point on the horizon, reaches a maximum altitude above the south point of 90° minus your terrestrial latitude (which we'll write as $\bar{\varphi} = 90° - \varphi$), and falls back down to the west point. The armillary sphere also contains a circle that crosses the celestial equator at an angle of $\varepsilon = 23.4°$, equal to the tilt of the Earth's axis. This circle, the *ecliptic*, is the path traversed by the Sun through the celestial sphere over the course of the year, crossing the equator at the *vernal* (or *spring*) *equinox*, ♈ in Figure 69. Since the year is roughly 365 days long and there are 360 degrees in a circle, the Sun moves about 1° per day along the ecliptic. The proximity of these two numbers is not a coincidence; the ancient Babylonian astronomers chose to divide the circle into 360 degrees so that the Sun would move about a degree per day.

In Figure 69 the Sun has moved about $\lambda = 60°$ beyond ♈, so it is late May, about sixty days after the vernal equinox. Unfortunately, this information doesn't help us locate the Sun in the sky; over the course of the day the ecliptic changes its position appreciably. Therefore it is customary to pinpoint the Sun's location with respect to the equator, which stays in place. Drop a perpendicular on the celestial sphere from the Sun to point J on the equator. The Sun's *equatorial coordinates* are the *right ascension* α (measured along the equator) and the *declination* δ (measured

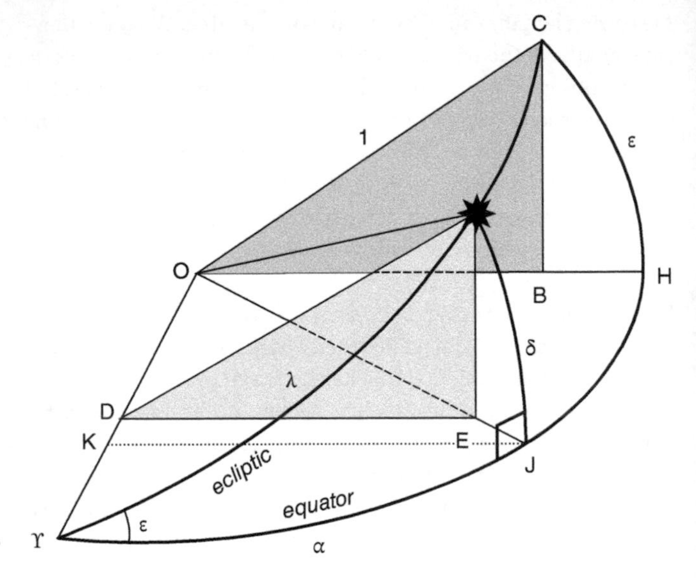

69. Calculating the declination and right ascension of an arc of the ecliptic.

at right angles to the equator). We know λ and ε; our goal is to find δ and α.

We begin by connecting the sphere's centre *O* to various points on the sphere. Next we form the two shaded similar right triangles *OBC* and *DE*✷ by dropping perpendiculars from *C* and ✷ onto the plane of the equator, then dropping perpendiculars from the resulting points *B* and *E* onto *O*ϒ. What are the angles in these triangles at *O* and *D*? Imagine tilting the front of the diagram upwards so that we view it from ϒ, with *O* directly behind. From this perspective, the ecliptic coincides with *OC* and *D*✷, while the equator coincides with *OB* and *DE*. Therefore the angle ε between the ecliptic and the equator is equal to the angles at *O* and *D*.

We're now in a position to do some trigonometry. In triangle *OBC* we know that $OC = 1$ and $\measuredangle COB = \varepsilon$, so $BC = \sin\varepsilon$. Next, we turn to triangle ✸*OD*. Since $\measuredangle$✸*O*♈ is equal to arc ✸♈ $= \lambda$, and ✸*O* is equal to 1, we know that ✸*D* $= \sin\lambda$. Finally, consider triangle *EO*✸. Since $\measuredangle$*JO*✸ is equal to arc *J*✸ $= \delta$ and *O*✸ $= 1$, we know that *E*✸ $= \sin\delta$.

We now invoke the similarity of the two triangles and put our pieces together. Since

$$\frac{BC}{OC} = \frac{E✸}{D✸},$$

we have

$$\frac{\sin\varepsilon}{1} = \frac{\sin\delta}{\sin\lambda},$$

which gives us the standard formula for the declination:

$$\sin\delta = \sin\varepsilon \sin\lambda.$$

We used this formula back in Chapter 4, but now we know that it's true. From $\lambda = 60°$ and $\varepsilon = 23.4°$, we calculate $\delta = 20.12°$.

Now that we have the declination δ, we can find the right ascension α. We'll leave this as a puzzle, with a couple of hints. In Figure 69, draw *JK* perpendicular to *O*♈. This time our similar right triangles are *OED* and *OJK*. The first step is to recognize from triangle *DO*✸ that $OD = \cos\lambda$. The rest is up to you. The conclusion is

$$\cos\lambda = \cos\delta \cos\alpha.$$

Since we already know λ and δ, this formula allows us to compute α; we get $\alpha = 57.83°$.

The ten formulas for right-angled spherical triangles

We can think of Figure 69 as an astronomical diagram, but we can also ignore its astronomical content and think entirely geometrically. Relabel the vertices of spherical right triangle Υ✱J by calling the vertices A, B, and C (where C is the right angle), and call the sides opposite the corresponding angles a, b, and c (Figure 70). Our declination and right ascension formulas become

$$\sin a = \sin A \sin c$$

and

$$\cos c = \cos a \cos b.$$

This latter formula is especially interesting, because it plays the same role with spherical right triangles that $c^2 = a^2 + b^2$ plays with planar right triangles: it allows us to find the hypotenuse, given the lengths of the other two sides. In other words, we have the

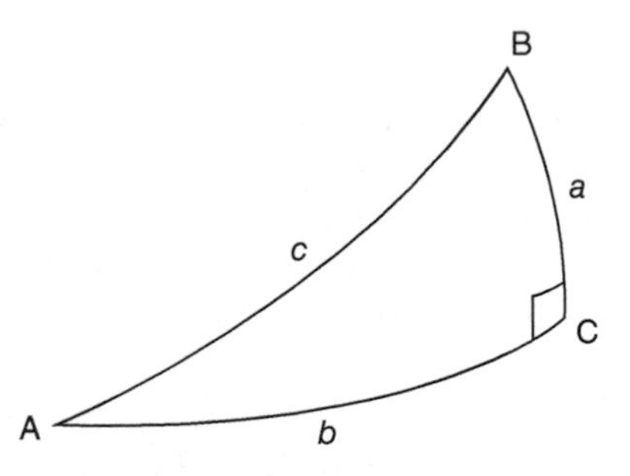

70. A spherical right triangle.

SPHERICAL PYTHAGOREAN THEOREM: $\cos c = \cos a \cos b$.

Considering that a spherical triangle appears to be more complicated than a plane triangle, the simplicity of this formula is astonishing. And although it doesn't look much like $c^2 = a^2 + b^2$, there is a connection. Recall that a tiny spherical triangle drawn on a large sphere is almost a plane triangle, and also (from Chapter 5) that

$$\cos\theta = 1 - \frac{\theta^2}{2!} + \frac{\theta^4}{4!} - \frac{\theta^6}{6!} + \frac{\theta^8}{8!} - \frac{\theta^{10}}{10!} + \frac{\theta^{12}}{12!} - \ldots$$

If θ is very small, then each term is much smaller than the previous one, and we can approximate the cosine with just the first couple of terms:

$$\cos\theta \approx 1 - \frac{\theta^2}{2!}.$$

For our tiny spherical triangle, the spherical Pythagorean theorem becomes

$$1 - \frac{c^2}{2!} \approx \left(1 - \frac{a^2}{2!}\right)\left(1 - \frac{b^2}{2!}\right).$$

Cleaning up the algebra, we get

$$c^2 \approx a^2 + b^2 - \frac{a^2 b^2}{2}.$$

But if a and b are tiny, then $a^2 b^2 / 2$ is much tinier than the other terms. Removing it leaves us with $c^2 \approx a^2 + b^2$.

The two formulas for right spherical triangles that we've found so far are only the tip of the iceberg. With not much more effort, it's possible to add more and more formulas to the list. Altogether, there are ten to be found:

$$
\begin{array}{ll}
\sin b = \tan a \cot A & \sin a = \sin A \sin c \\
\cos c = \cot A \cot B & \cos A = \sin B \cos a \\
\sin a = \cot B \tan b & \cos B = \cos b \sin A \\
\cos A = \tan b \cot c & \sin b = \sin c \sin B \\
\cos B = \cot c \tan a & \cos c = \cos a \cos b
\end{array}
$$

As far as we know, the first person to find and gather together these ten identities was probably Georg Rheticus (1514–74). Rheticus is best known as Copernicus' only student, the person who finally convinced Copernicus to publish his theory of the Sun-centred solar system. But he was especially interested in trigonometry; he was the first European to identify and tabulate all six of the standard trigonometric functions. The first to publish the ten identities in a list was François Viète (1540–1603), whose name is associated with the invention of symbolic algebra. But neither of these two luminaries seems to have recognized the amazing symmetries hidden in these identities. You may wish to inspect them again to see what you can find before moving on.

The first pattern to notice is that each identity in the left column consists of a co/sine equal to a co/tangent multiplied by a co/tangent, while each identity in the right column contains only sines and cosines. But there is much more. We can get a hint by looking down the *columns*. Reading the variables in the left side of the equal sign in the left column, we have b, c, a, A, and B. On the other side of the equal sign, read the variables downwards starting at the $\tan b$ in the fourth row (looping back to the top once we reach the bottom): b, c, a, A, and B. In each of the six columns of variables in the table, we read the same sequence: b, c, a, A, and B.

The symmetries were noticed by John Napier in his 1614 *Mirifici logarithmorum canonis descriptio*, the same book in which he

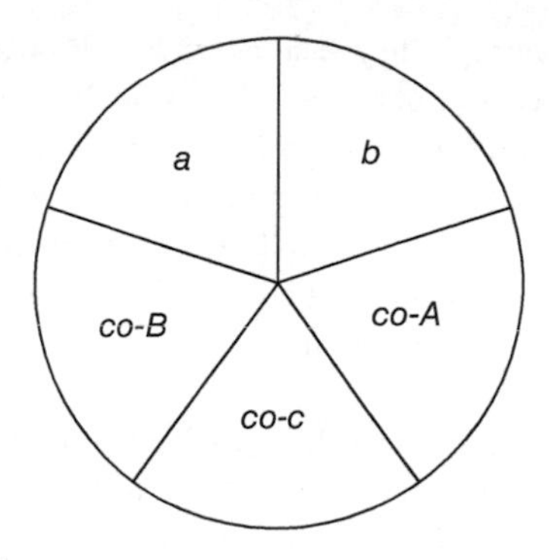

71. Napier's Rules.

announced his discovery of logarithms. They are encapsulated as follows, using Figure 71:

NAPIER'S RULES

(1) The sine of any circular part is equal to the product of the tangents of the two parts adjacent to it.
(2) The sine of any circular part is equal to the product of the cosines of the two parts opposite to it.

The term 'circular part' refers to any of the slices in the diagram; the 'co-' instructs us to switch from sine/tangent to cosine/cotangent or vice versa. For instance, if we apply Rule 1 to the slice 'co-A' we get $\cos A = \tan b \cot c$, the fourth identity in the left column.

The pentagramma mirificum

The pedagogical community reacted to Napier's rules extremely positively; over the years various physical devices based on Napier's rules were invented to help recall the ten identities. Scholars, having memorized the identities already, were not so kind. The famous 19th-century English logician Augustus DeMorgan claimed that they 'only create confusion instead of assisting the memory'. They seem not to have been aware that the symmetries reflect one of the most beautiful theorems in all of mathematics. John Napier did; it is in his book on logarithms. However, in an all too common manoeuvre (especially today), textbooks eventually disposed of it. Although the theorem explains the symmetries marvellously, it does not help a student solve a triangle during an examination.

Thankfully here we are not subject to the tyrannies of examination preparation, so we will take a closer look. Before we begin, we'll need one simple fact. If a spherical triangle has two sides equal to 90° (Figure 72), then the third side meets these two sides at right angles, and the angle θ where the two sides meet is equal in magnitude to the third side.

Turning now to our theorem: in Figure 73 the right-angled triangle has been placed at the top of the diagram, and the three sides extended into longer arcs. Using A as a pole, draw equator *UWVS*; and using B as a pole, draw equator *RXWT*. This gives us the *pentagramma mirificum*, or 'miraculous pentagram'. (If you trust me, feel free to jump ahead a couple of paragraphs to Figure 74 where the values of all the arcs and angles are labelled.) First, notice that all five of the triangles that form the 'petals' of the pentagram are right angled; we know this because each of the vertices S, U, R, and T are places where an arc travelling from a pole reaches its equator. Next, consider arc ABS. Since it travels from pole A to equator US, we know its length is 90°. But $AB = c$,

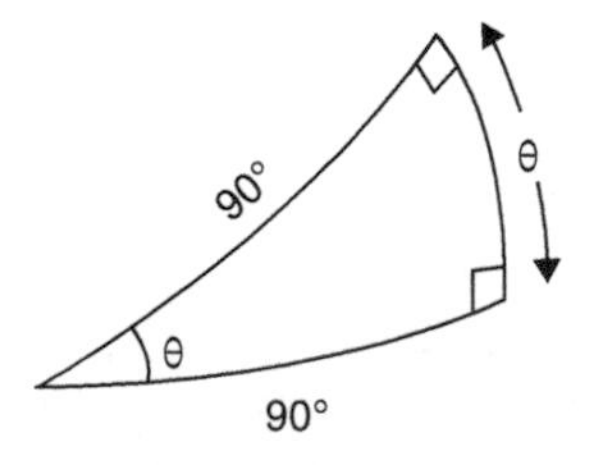

72. A simple fact about certain spherical triangles.

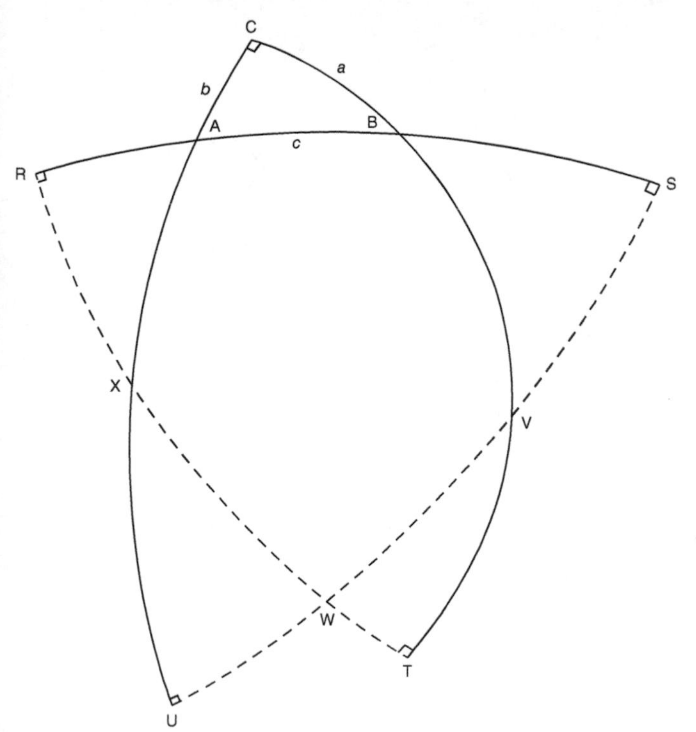

73. The *pentagramma mirificum*. The dashed arcs are equators corresponding to poles *A* and *B*.

so $BS = 90° - c$, which we call $\overline{c}$. With this sort of logic, we can find that *any* two adjacent arcs sum to 90°. This allows us to determine the lengths of all the arcs on *RS*, *CU*, and *CT*.

We turn next to the arcs on *US* and *RT*. Consider triangle *AUS*. Two of its angles are 90°; the third angle is $180° - A$. From Figure 72, we know that $US = 180° - A$. But $UV = 90°$, so $SV = 180° - A - 90° = 90° - A = \overline{A}$. We can do the same thing with arc *RT* and triangle *BRT*. At this point we can find all the arcs in the diagram. Finally, we turn to the missing angles in the figure.

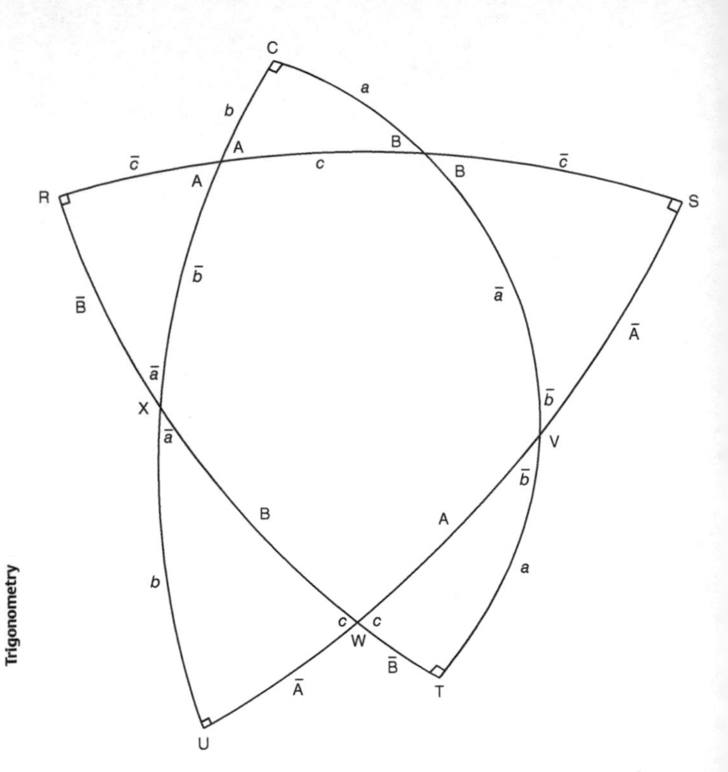

74. The *pentagramma mirificum* with the known arcs and angles filled in.

Consider triangle *WRS*. The three arcs that make up *RS* add up to $180° - c$. By Figure 72, this implies that $\measuredangle XWV = 180° - c$. But $\measuredangle XWV + \measuredangle XWU = 180°$, so $\measuredangle XWU = c$. We can determine the rest of the angles in a similar way. Now we know all the angles and arcs, which we write on Figure 74.

We can now use the *pentagramma* to generate all ten identities from only the top two. Take the identity at the top of the right column, $\sin a = \sin A \sin c$. Move from the top triangle *ABC* to the one to its right, *BSV*. What was side a is now $\overline{A} = 90° - A$; what was A is now B; and what was c is now $\bar{a}$. If we apply the identity to BSV, we get $\sin \bar{A} = \sin B \sin \bar{a}$, which is the same as $\cos A = \sin B \cos a$. This is the second identity in the right column! Do the same thing to the third triangle VTW, and we get the third identity. Likewise for the fourth and fifth. The same pattern holds for the left column: take the identity at the top, apply it one by one to the other four triangles, and the entire column appears as if by magic. Truly, the pentagram has earned the name *mirificum*.

Oblique triangles

When we learned plane trigonometry, we began with right triangles and moved on to oblique triangles. The same is true in spherical trigonometry. For example: as I was flying from Vancouver to Edmonton I spent some time watching the on-board navigation screen in the flight entertainment system—for a spherical trigonometer, much more interesting than the latest Hollywood blockbuster. Connect both Vancouver and Edmonton to the North Pole, forming an oblique spherical triangle (Figure 75). Our heading when we left Vancouver was 50.7° east of north; when we arrived our heading had changed to 58.22° east of north, so the angle at Edmonton in the diagram is $180° - 58.22° = 121.78°$. When we left Vancouver the screen told me that our latitude was 49.3° north; when we arrived in Edmonton I forgot to check the latitude. Can we reconstruct it?

The latitudes of Vancouver and Edmonton are 90° minus their respective distances to the North Pole, so $AC = 90° - 49.3° = 40.7°$; we need to find BC. We drop a perpendicular from the North Pole onto our journey AB, splitting our oblique triangle into two right-angled triangles. We can apply the identity $\sin a = \sin A \sin c$ to get an equation for the length h, using either the left triangle or the right. We get

$$\sin h = \sin A \sin b \quad \text{and} \quad \sin h = \sin B \sin a.$$

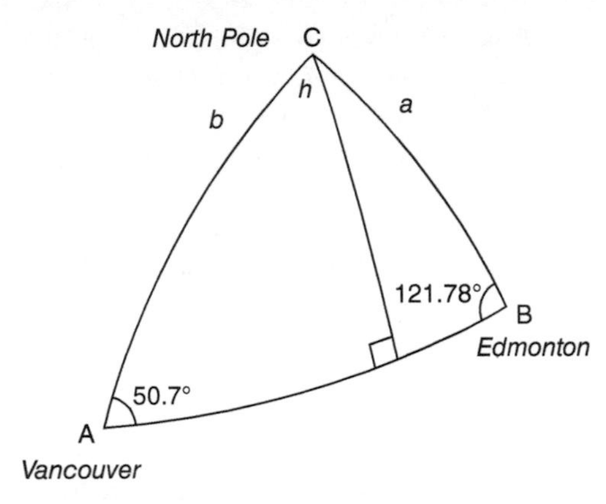

75. Travelling from Vancouver to Edmonton.

Both of these formulas include $\sin h$, so we can set them equal to each other and save ourselves from computing h at all, which we really didn't care about in any case. So $\sin A \sin b = \sin B \sin a$. Rearranging and including c and C (since we could have applied this argument from any vertex), we have the

$$\text{SPHERICAL LAW OF SINES: } \frac{\sin a}{\sin A} = \frac{\sin b}{\sin B} = \frac{\sin c}{\sin C}.$$

This pleasingly symmetric formula looks similar to its planar counterpart:

$$\text{PLANAR LAW OF SINES: } \frac{a}{\sin A} = \frac{b}{\sin B} = \frac{c}{\sin C}.$$

The spherical law gives us $\sin BC = \sin AC \cdot \sin A / \sin B$, and from this we find $BC = 36.4°$. Thus Edmonton's latitude is $90° - 36.4° = 53.6°$.

A natural question arises. If there is a spherical Law of Sines, is there also a spherical Law of Cosines? Although space precludes us from going into detail here, the answer is 'yes, and then yes again'. Recall the Law of Cosines from Chapter 4:

$$\text{PLANAR LAW OF COSINES: } c^2 = a^2 + b^2 - 2ab\cos C.$$

The spherical law has a similar structure:

$$\text{SPHERICAL LAW OF COSINES:}$$
$$\cos c = \cos a \cos b + \sin a \sin b \cos C.$$

In both cases, the Law of Cosines begins with a Pythagorean theorem, and then appends a correction term to account for the fact that C is not necessarily equal to 90°. Why 'yes, and then yes again'? It turns out that there is a *second* spherical Law of Cosines, this one concentrating on angles rather than sides:

$$\text{SPHERICAL LAW OF COSINES FOR ANGLES:}$$
$$\cos C = -\cos A \cos B + \sin A \sin B \cos c.$$

Navigating by the stars

Spherical trigonometry was intended originally for astronomers, but we've already seen some situations where it aided those with Earth-bound concerns. The earliest occasions were provided by medieval Islamic scholars who used it to resolve demands required by ritual. They predicted the beginning of the sacred month of Ramadan, defined by the emergence of the lunar crescent from the glare of the Sun at the time of new Moon. They also determined the times of the five daily prayers, some of which require knowledge of the Sun's altitude. Finally, they found the direction of Mecca, which worshippers needed to face to pray.

There are also much more modern episodes of spherical trigonometry coming to the rescue. One of the most dramatic stories in the history of mathematics is the 1837 adventure of

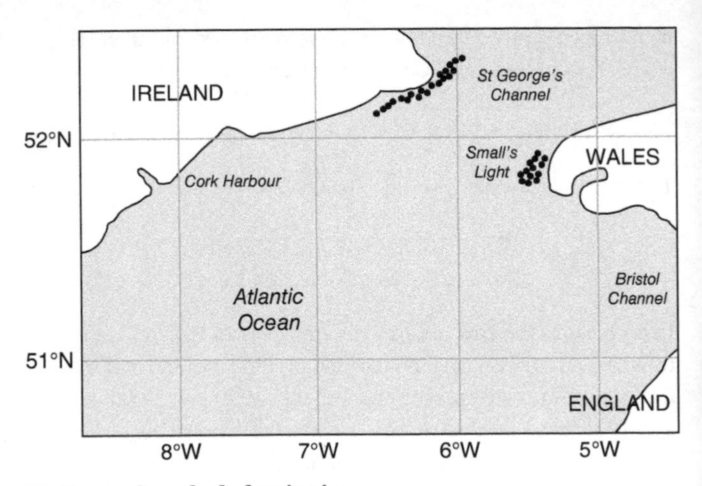

76. Sumner's method of navigation.

Captain Thomas Hubbard Sumner. He set off from South Carolina into the Atlantic Ocean, and three weeks later he needed to sail through St George's Channel between Wales and Ireland on his way to Scotland. However, miserable weather and obscured skies made him unsure of his position, and potentially fatal rocks awaited along the south shore of Ireland (Figure 76). The clouds parted momentarily, which gave him just enough time to measure the Sun's altitude, 12°10′ above the horizon. Then, his creativity perhaps sharpened by the stakes of survival, he reasoned as follows. The collection of places on the Earth's surface where the Sun is at a given altitude forms a circle whose centre is the Sun's geographic position (GP)—that is, the point on the Earth directly below the Sun (Figure 77). Sumner knew he had to be somewhere on that circle, and he could calculate its position.

Such a circle is called a *small circle*, not because it is small, but because it isn't a great circle. Sumner's circle was actually extremely large—so large that the part of it on our map is almost straight, called the *line of position*. By great fortune, Sumner's line of position happened to pass through the sea in a north-easterly

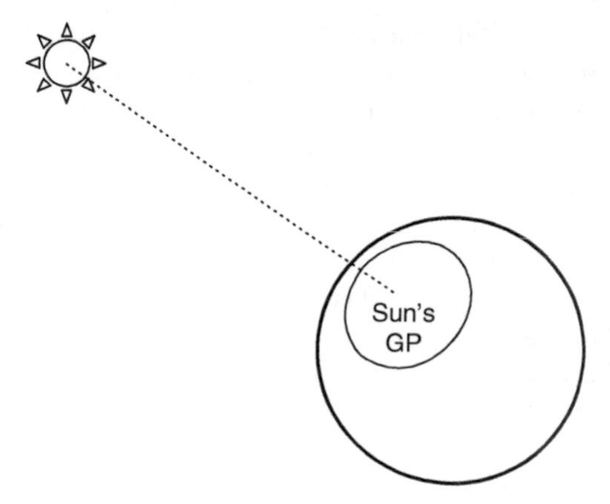

77. The Sun's geographic position (GP). At every point on the circle around the Sun's GP, the Sun is seen at the same altitude.

direction (roughly between 51° N, 8° W and 52° N, 5° W in Figure 76) and very nearly contacted Smalls Lighthouse off the coast of Wales in a well-charted region. Although Sumner didn't know where on the line he was, all he had to do was keep travelling along it. He would be assured of spotting Smalls Lighthouse eventually, and from there he could navigate to safety.

An extension of Sumner's ingenious reasoning allows you to pinpoint your location on Earth, no matter where you are. If measuring the altitude of one star places you on some small circle on the Earth's surface, measuring the altitude of two stars places you at the intersection of a pair of circles. These circles intersect at two points, one of which is your location. Almost always, those two points are very far away from each other, and if you cannot tell whether you're off the south coast of Ireland or the south coast of India, you have bigger problems than navigation can solve. In practice, you can determine a ship's location in this way to within a kilometre, and the reliability of the method can be increased by observing the altitudes of more than two stars.

Modern technologies such as GPS have rendered traditional practices of spherical trigonometry obsolete other than for hobbyists. But they are making at least a small comeback. At the US Naval Academy in Annapolis, Maryland, one of the last institutions to give up teaching spherical trigonometry decades ago, officers in training are once again being instructed in celestial navigation. The potential for GPS systems to be jammed by enemies at times of conflict may cause threatened sailors to turn their eyes to the heavens, not to cry for help from divine powers, but to apply ancient ingenuity to save themselves and their shipmates.

Beyond Euclid

We have already encountered the perhaps the most important book in the history of mathematics, Euclid's *Elements*. It contained not much that was new, even in its own time. The book's greatness came from *how* Euclid presented the material. He demonstrated each theorem logically, each one from those he had already established. Of course, to prove any theorem you need to start somewhere. So at the beginning of his book he listed five postulates and five common notions, today called *axioms*, that he didn't prove at all. The mark of a good axiom is that it is

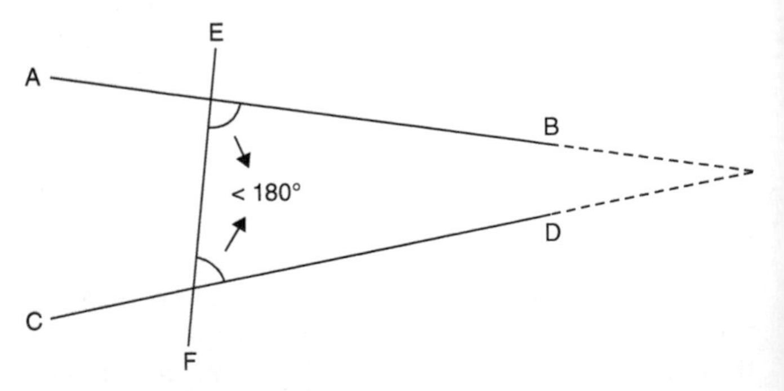

78. Euclid's parallel postulate.

self-evident (that is, transparently true to anyone) and *simple*. If one is going to make an assumption at the foundation of a logical process, one had better hope that it is not questionable: any level of doubt about any axiom permeates the entire system that is built upon it.

Some of Euclid's axioms are:

- All right angles are equal to each other.
- Things which are equal to the same thing are equal to each other.
- If equals are added to equals, then the wholes are equal.

It's hard (most would say impossible) to imagine how these simple statements could possibly be false. If any of these statements were not to be true, one can hardly imagine doing mathematics at all. However, one of Euclid's axioms, the *parallel postulate*, was not like the others:

> If a straight line falling on two straight lines makes the interior angles on the same side less than two right angles, the two straight lines, if produced indefinitely, meet on that side on which are the angles less than the two right angles.

It takes a little effort even to understand what the statement is saying, let alone judge whether it's true or self-evident. In Figure 78, the two original straight lines are *AB* and *CD*; the straight line falling on them is *EF*. According to the parallel postulate, if the two angles drawn in the figure sum to less than two right angles (<180°), then *AB* and *CD* extended must eventually meet on the right side of *EF*.

Now that you see it on paper, you are probably much more willing to concede that the postulate is true, even if it is not as simple as the other axioms. None of Euclid's colleagues or successors for two millennia doubted its truth, although they were unhappy with it. Euclid himself disliked it; he avoided using it as long as he possibly could. Through the 18th century, various geometers took on the parallel postulate as a challenge: either to prove it from the other axioms, thereby freeing it from 'axiom' status; or to replace it with another axiom that is simpler and therefore easier to accept as a starting point. Two axioms that one might use as an alternative to the parallel postulate are:

- Given a line and a point not on it, there is precisely one line that passes through the point but never touches the line.
- The angles in a triangle sum to 180°.

These two statements and the parallel postulate are all logically equivalent: that is, it can be proved that either all three are true, or all three are false.

If you are encountering these postulates for the first time, likely you don't even consider the possibility that they're all false. Curiously, a counterexample was available already in Euclid's time, in a discipline on which Euclid actually wrote: spherical geometry. As we have seen, if you walk in a straight line on the surface of a sphere, you form a great circle. There is no such thing as a pair of parallel great circles; any two of them intersect at two antipodal points. As we've also seen, the sum of the angles in a spherical triangle is greater than 180°.

Euclid and his colleagues did not consider this to be a counterexample any more than you probably do. A straight line is a straight line, and as is clear when we look at a great circle from outside the sphere, a great circle is not straight. But if we replace the word 'line' in Euclid's axioms with the words 'great circle', all the axioms hold true—all, that is, except the parallel postulate. If there were to be a proof of the parallel postulate for ordinary lines from the other axioms, then we could just replace 'line' with 'great circle' and have a proof for great circles. However, the postulate is false for great circles. Therefore, we know that the parallel postulate can never be proved from the other axioms.

Efforts to resolve this logical annoyance suddenly transformed into a much larger conundrum in the 19th century through the work of three mathematicians working independently: Carl Friedrich Gauss (1777–1855), János Bolyai (1803–60), and Nikolai Lobachevsky (1792–1856). All three considered the implications of accepting that the parallel postulate is *false*. How might this be? When we look at a great circle from outside the sphere, it does not look straight to us; but to someone whose entire universe is the sphere's surface, it is perfectly straight. Who's right? Who's to say? Imagine that we are sphere surface dwellers whose entire experience is in such a tiny region of the sphere that the surface seems to be perfectly flat. We draw triangles, and their angles appear to add up to 180°, although in reality the sums are very, very slightly greater. We humans live in such a tiny corner of our universe that we might as well be the sphere surface dwellers. We do not know even today whether the angles of a triangle in our universe sum to 180°. Gauss, Bolyai, and Lobachevsky assumed that they do not, and followed the implications of that claim. While they found stranger and stranger facts, none were logically impossible—just outside of our limited experience. Reflecting on his bizarre journey, Bolyai marvelled, 'out of nothing I have created a strange new universe'.

It turns out there are two types of these so-called *non-Euclidean geometries*. In *elliptical geometry* (of which the sphere is a variant), triangles' angles sum to greater than 180° and the space is said to have positive curvature. In *hyperbolic geometry*, triangles' angles sum to less than 180° and the space is said to have negative curvature. In elliptical geometry there are no lines through a given point parallel to a given line; in hyperbolic geometry there are infinitely many.

Non-Euclidean universes can be very difficult to grasp without something visual to hold on to, such as the sphere for elliptical geometry. Fortunately, there are several models for hyperbolic

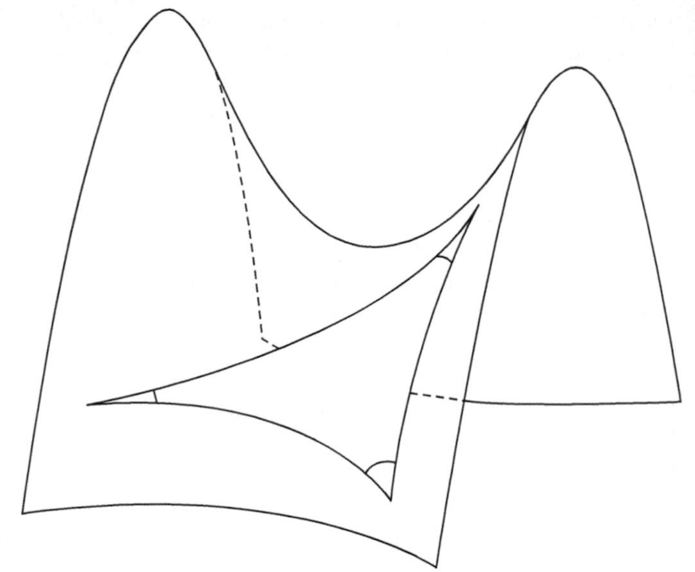

79. A model of hyperbolic geometry. In this space the angles of a triangle sum to less than 180°.

geometry. The first model explains the name 'hyperbolic': simply replace the sphere with a *hyperbolic paraboloid* (Figure 79). As one can see, the sides of a triangle on the surface of this figure bend towards each other at the vertices, so the sum of the angles is less than 180°. Another model is the *Poincaré disc* (Figure 80), the interior of a unit circle. In this model the 'straight lines' are arcs of circles that meet the edge of the unit circle at right angles. Distances within the Poincaré disc are also different from what we see from the outside. If we were to walk towards the boundary we would never reach it; from within the disc, distances are stretched more and more as we get closer to the edge. The outer boundary, in fact, is infinitely far away. If the distance from the disc's centre to a given point is x as perceived from outside the disc, then the distance perceived within the disc turns out to be $2\tanh^{-1} x$, a quantity that goes to infinity as x approaches 1.

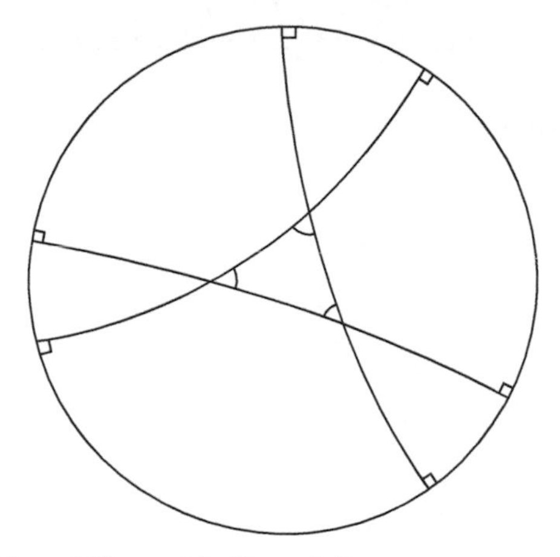

80. The Poincaré disc model of hyperbolic geometry. The 'straight lines' of this space are circular arcs that touch the boundary circle at right angles. The angles of a triangle sum to less than 180°.

The appearance of $\tanh^{-1}$ is our first hint of the unsurprising fact that the hyperbolic trigonometric functions have a major role to play in hyperbolic trigonometry. The parallels between spherical and hyperbolic trigonometry are almost perfect; to convert almost any formula in spherical trigonometry to one in hyperbolic trigonometry, one needs only to convert the functions whose arguments are lengths (not angles) into their hyperbolic equivalents. For instance, here are the ten standard identities for right hyperbolic triangles:

$$\sinh b = \tanh a \cot A \qquad \sinh a = \sin A \sinh c$$
$$\cosh c = \cot A \cot B \qquad \cos A = \sin B \cosh a$$
$$\sinh a = \cot B \tanh b \qquad \cos B = \cosh b \sin A$$
$$\cos A = \tanh b \coth c \qquad \sinh b = \sinh c \sin B$$
$$\cos B = \coth c \tanh a \qquad \cosh c = \cosh a \cosh b$$

The *pentagramma mirificum* appears to exist not only on the sphere, but also in hyperbolic space!

The parallels continue when we turn to oblique triangles.

HYPERBOLIC LAW OF SINES: $\frac{\sinh a}{\sin A}=\frac{\sinh b}{\sin B}=\frac{\sinh c}{\sin C}$

HYPERBOLIC LAW OF COSINES:
$$\cosh c = \cosh a \cosh b - \sinh a \sinh b \cos C$$

HYPERBOLIC LAW OF COSINES FOR ANGLES:
$$\cos C = -\cos A \cos B + \sin A \sin B \cosh c$$

Notice that the parallel with spherical geometry is broken here in one minor way: the right side of the hyperbolic Law of Cosines has a minus sign on the right side of the equation, whereas the spherical law had a plus sign.

When Gauss, Bolyai, and Lobachevsky were exploring their strange new universes, they had no physical manifestation of their theories in mind; they were simply new intellectual constructs that challenged basic assumptions of how geometry should work. Less than a century later, hyperbolic trigonometry found a home as part of Einstein's special relativity. Although great mathematics often follows from great science, every once in a while the great mathematics comes first.

Our *Very Short Introduction* to trigonometry is at an end, but there are many ways that one can continue the adventure. See Further Reading for sources that expand upon the topics we've seen here. From its beginning as a merger of geometry and computation to track the motions of the heavens in ancient Greece, trigonometry has gone on to interact with some of the most important developments in science. On its own, it has created some of the world's most beautiful mathematics. The small corner of trigonometry found in textbooks is only the tip of an astonishing iceberg.

Further reading

Chapter 1: Why?

There are many books from which you can learn the history of astronomy, either briefly or in detail. These books are at either end of the range.

Michael Hoskin, *The History of Astronomy: A Very Short Introduction*. Oxford: Oxford University Press, 2003.

John North, *Cosmos: An Illustrated History of Astronomy and Cosmology*. Chicago: University of Chicago Press, 2008.

For a look specifically at mathematics in astronomy, see

C. M. Linton, *From Eudoxus to Einstein: A History of Mathematical Astronomy*. Cambridge: Cambridge University Press, 2004.

To learn ancient astronomy by actually doing it, there is no better source than

James Evans, *The History and Practice of Ancient Astronomy*. Oxford: Oxford University Press, 1998.

Chapter 2: Sines, cosines, and their relatives

This chapter contains much of the trigonometry learned in school, so almost any textbook will fill in some of the details and will provide exercises for practice. If you want to learn in more depth, there are few places as good as 19th-century textbooks. In trigonometry, consider the following by the most famous mathematics textbook author of his time, available on Google Books:

Isaac Todhunter, *Plane Trigonometry for the Use of Colleges and Schools*. 2nd edition. Cambridge: Macmillan, 1860.

To learn about the origins of the symbols and words in mathematics, this classic resource is still in print:

Florian Cajori, *A History of Mathematical Notations*, 2 vols. La Salle, IL: Open Court, 1928/9. Reprinted in one volume, New York: Dover, 1993.

For a more recent book, see

Joseph Mazur, *Enlightening Symbols: A Short History of Mathematical Notation and its Hidden Powers*. Princeton, NJ: Princeton University Press, 2014.

A pair of web sites operated by Jeff Miller is a good reference source on the history of mathematical notations:

Earliest Known Uses of Some of the Words of Mathematics, http://jeff560.tripod.com/mathword.html.

Earliest Uses of Various Mathematical Symbols, http://jeff560.tripod.com/mathsym.html.

The story of the birth of the tangent function in Europe is revealed in

Glen Van Brummelen, The end of an error: Bianchini, Regiomontanus, and the tabulation of stellar coordinates, *Archive for History of Exact Sciences* 72 (2018), 547–63.

On Jamshīd al-Kāshī's calculation of π as well as background on Archimedes' method, see

Glen Van Brummelen, Jamshīd al-Kāshī: Calculating genius, *Mathematics in School*, 27 (4) (1998), 40–4.

The original article on al-Samaw'al's discussion of the dip angle to the horizon is

J. L. Berggren and Glen Van Brummelen, Al-Samaw'al versus al-Kūhī on the depression of the horizon, *Centaurus* 45 (2003), 116–29.

Chapter 3: Building a sine table with your bare hands

To learn more about the processes involved in building a sine table from scratch according to ancient methods, see the first chapter of this book, available for free at the publisher's web site:

Glen Van Brummelen, *Heavenly Mathematics: The Forgotten Art of Spherical Trigonometry*. Princeton, NJ: Princeton University Press, 2013.

Āryabhaṭa's work on sine differences has been interpreted in at least fourteen different ways. The following article summarizes thirteen of them and presents a new one:

Takao Hayashi, Āryabhaṭa's rule and table for sine-differences, *Historia Mathematica* 24 (1997), 396–406.

Chapter 4: Identities, and more identities

The origin of the Law of Cosines is a thorny topic, discussed in

Glen Van Brummelen, Filling in the short blanks: Musings on bringing the historiography to the classroom, *BSHM Bulletin* 25 (2010), 2–9.

Mollweide's formula is not well known today. Here is an article on its use in the classroom, and two straightforward visual proofs:

Natanael Karjanto, Mollweide's formula in teaching trigonometry, *Teaching Mathematics and its Applications* 30 (2) (2011), 70–4.

H. A. DeKleine, Proof without words: Mollweide's equation, *Mathematics Magazine* 61 (1988), 281.

R. H. Wu, Proof without words: Mollweide's equation, *College Mathematics Journal* 32 (2001), 68–9.

The use of the sum-to-product formulas to describe explain the beat phenomenon in music may be found in section 1.8 (pp. 23–6) of

David Benson, *Music: A Mathematical Offering*. Cambridge: Cambridge University Press, 2006.

Morrie's Law as recalled by Richard Feynman is described on page 47 of

James Gleick, *Genius: The Life and Science of Richard Feynman*. New York: Pantheon, 1992.

Here are two articles on Morrie's Law. Our geometric extension of the law is based on the proof in the second article.

W. A. Beyer, J. D. Louck, and D. Zeilberger, A generalization of a curiosity that Feynman remembered all his life, *Mathematics Magazine* 69 (1996), 43–4.

Samuel G. Moreno and Esther M. García-Caballero, A geometric proof of Morrie's Law, *American Mathematical Monthly* 122 (2015), 168.

Chapter 5: To infinity...

The argument that leads to the Taylor series for the sine without using calculus comes from

John Quintanilla, The Taylor polynomials of sin θ, *College Mathematics Journal* 38 (2007), 58–9.

There are a number of accessible books dealing with infinity in mathematics; in fact, an entire category at Amazon is devoted to them. Among them are

David Foster Wallace, *Everything and More: A Compact History of Infinity*. New York and London: W. W. Norton, 2003.

John D. Barrow, *The Infinite Book*. New York: Vintage, 2005.
The first appearance of the Taylor series for the sine was with Mādhava, in 14th-century Kerala. An account of the argument by Jyeṣṭhadeva, a 16th-century member of Mādhava's school, is summarized on pages 113–21 of
Glen Van Brummelen, *The Mathematics of the Heavens and the Earth: The Early History of Trigonometry*. Princeton, NJ: Princeton University Press, 2009.
Mādhava's improvements of the infinite series for π are described in
Takao Hayashi, T. Kusuba, and Michio Yano, The correction of the Mādhava series for the circumference of a circle, *Centaurus* 33 (1990), 149–74.
A description of the CORDIC algorithm, and the story of its invention by the inventor, are here:
Alan Sultan, CORDIC: How hand calculators calculate, *College Mathematics Journal* 40 (2009), 87–92.
Jack E. Volder, The birth of CORDIC, *Journal of VLSI Signal Processing* 25 (2000), 101–5.
On Thomson (Lord Kelvin) and his work on tides and Fourier series, there are several places to go. Thomson's original article on the subject is available at https:// HathiTrust.org:
Sir William Thomson, Baron Kelvin, The tide gauge, tidal harmonic analyser, and tide predicter, in *Mathematical and Physical Papers*, vol. 6 (Cambridge: Cambridge University Press, 1911), pp. 272–305.
Two good online sources for more detail on Thomson's devices and Fourier series are:
Tony Phillips, Fourier analysis of ocean tides (3 parts), American Mathematical Society Feature Column, http://www.ams.org/publicoutreach/feature-column/fcarc-tidesi1.
Raymond Flood, Gresham College lecture 'Fourier's Series', https://www.youtube.com/watch?v=3bQz6k2FRu4.
See also the article
Robert K. Otnes, Notes on mechanical Fourier analyzers, *Journal of the Oughtred Society* 17 (2008), 34–41.

Chapter 6: ... and beyond, to complex things

Euler's formula $e^{i\pi}+1=0$ has been discussed and celebrated many times. A lovely recent book on the formula is
Robin Wilson, *Euler's Pioneering Equation*. Oxford: Oxford University Press, 2018.

A good book on the story of imaginary and complex numbers is

Paul Nahin, *An Imaginary Tale: The Story of* $\sqrt{(-1)}$. Princeton, NJ: Princeton University Press, 1998.

The complicated history of the birth and emergence of the hyperbolic functions is recounted in

Janet Heine Barnett, Enter, stage center: The early drama of the hyperbolic functions, *Mathematics Magazine* 77 (2004), 15–30.

Several derivations of the hyperbolic cosine as the solution to the catenary problem are available online. For instance, see https://www.math24.net/equation-catenary/.

Chapter 7: Spheres and more

Part of this chapter is adapted with permission from Glen Van Brummelen, Trigonometry for the heavens, *Physics Today* 70 (2017), 70–1.

A modern entry into spherical trigonometry with historical overtones is

Glen Van Brummelen, *Heavenly Mathematics: The Forgotten Art of Spherical Trigonometry*. Princeton, NJ: Princeton University Press, 2013.

For a more rigorous treatment of the subject, one can go back to an old classic:

Isaac Todhunter and J. G. Leatham, *Spherical Trigonometry for the Use of Colleges and Schools*. London: Macmillan, 1901.

Or, one can go to a brand new book:

Marshall Whittlesey, *Spherical Geometry and its Applications*. Boca Raton, FL: CRC Press, 2019.

To read more about non-Euclidean geometry and its potential applications to physics and cosmology, see

Robert Osserman, *Poetry of the Universe*. New York: Anchor, 1995.

If you wish to sink your teeth into the mathematics of non-Euclidean geometry, consider the upper-level undergraduate textbook

Jeremy Gray, *Worlds Out of Nothing: A Course in the History of Geometry in the 19th Century*. London: Springer, 2007.

Index

A

B

C

I

J

K

L

M

N

CHAOS

A Very Short Introduction

Leonard Smith

Our growing understanding of Chaos Theory is having fascinating applications in the real world - from technology to global warming, politics, human behaviour, and even gambling on the stock market. Leonard Smith shows that we all have an intuitive understanding of chaotic systems. He uses accessible maths and physics (replacing complex equations with simple examples like pendulums, railway lines, and tossing coins) to explain the theory, and points to numerous examples in philosophy and literature (Edgar Allen Poe, Chang-Tzu, Arthur Conan Doyle) that illuminate the problems. The beauty of fractal patterns and their relation to chaos, as well as the history of chaos, and its uses in the real world and implications for the philosophy of science are all discussed in this *Very Short Introduction*.

> '. . . Chaos . . . will give you the clearest (but not too painful idea) of the maths involved . . . There's a lot packed into this little book, and for such a technical exploration it's surprisingly readable and enjoyable - I really wanted to keep turning the pages. Smith also has some excellent words of wisdom about common misunderstandings of chaos theory . . .'
>
> popularscience.co.uk

www.oup.com/vsi

Economics

A Very Short Introduction

Partha Dasgupta

Economics has the capacity to offer us deep insights into some of the most formidable problems of life, and offer solutions to them too. Combining a global approach with examples from everyday life, Partha Dasgupta describes the lives of two children who live very different lives in different parts of the world: in the Mid-West USA and in Ethiopia. He compares the obstacles facing them, and the processes that shape their lives, their families, and their futures. He shows how economics uncovers these processes, finds explanations for them, and how it forms policies and solutions.

> 'An excellent introduction . . . presents mathematical and statistical findings in straightforward prose.'
>
> Financial Times

INFORMATION

A Very Short Introduction

Luciano Floridi

Luciano Floridi, a philosopher of information, cuts across many subjects, from a brief look at the mathematical roots of information - its definition and measurement in 'bits'- to its role in genetics (we are information), and its social meaning and value. He ends by considering the ethics of information, including issues of ownership, privacy, and accessibility; copyright and open source. For those unfamiliar with its precise meaning and wide applicability as a philosophical concept, 'information' may seem a bland or mundane topic. Those who have studied some science or philosophy or sociology will already be aware of its centrality and richness. But for all readers, whether from the humanities or sciences, Floridi gives a fascinating and inspirational introduction to this most fundamental of ideas.

'Splendidly pellucid.'

Steven Poole, The Guardian

www.oup.com/vsi

NUMBERS
A Very Short Introduction
Peter M. Higgins

Numbers are integral to our everyday lives and feature in everything we do. In this *Very Short Introduction* Peter M. Higgins, the renowned mathematics writer unravels the world of numbers; demonstrating its richness, and providing a comprehensive view of the idea of the number. Higgins paints a picture of the number world, considering how the modern number system matured over centuries. Explaining the various number types and showing how they behave, he introduces key concepts such as integers, fractions, real numbers, and imaginary numbers. By approaching the topic in a non-technical way and emphasising the basic principles and interactions of numbers with mathematics and science, Higgins also demonstrates the practical interactions and modern applications, such as encryption of confidential data on the internet.

www.oup.com/vsi

STATISTICS

A Very Short Introduction

David J. Hand

Modern statistics is very different from the dry and dusty discipline of the popular imagination. In its place is an exciting subject which uses deep theory and powerful software tools to shed light and enable understanding. And it sheds this light on all aspects of our lives, enabling astronomers to explore the origins of the universe, archaeologists to investigate ancient civilisations, governments to understand how to benefit and improve society, and businesses to learn how best to provide goods and services. Aimed at readers with no prior mathematical knowledge, this *Very Short Introduction* explores and explains how statistics work, and how we can decipher them.

www.oup.com/vsi

RELATIVITY

A Very Short Introduction

Russell Stannard

100 years ago, Einstein's theory of relativity shattered the world of physics. Our comforting Newtonian ideas of space and time were replaced by bizarre and counterintuitive conclusions: if you move at high speed, time slows down, space squashes up and you get heavier; travel fast enough and you could weigh as much as a jumbo jet, be squashed thinner than a CD without feeling a thing - and live for ever. And that was just the Special Theory. With the General Theory came even stranger ideas of curved space-time, and changed our understanding of gravity and the cosmos. This authoritative and entertaining *Very Short Introduction* makes the theory of relativity accessible and understandable. Using very little mathematics, Russell Stannard explains the important concepts of relativity, from E=mc2 to black holes, and explores the theory's impact on science and on our understanding of the universe.

www.oup.com/vsi

GLOBALIZATION

A Very Short Introduction

Manfred Steger

'Globalization' has become one of the defining buzzwords of our time - a term that describes a variety of accelerating economic, political, cultural, ideological, and environmental processes that are rapidly altering our experience of the world. It is by its nature a dynamic topic - and this *Very Short Introduction* has been fully updated for 2009, to include developments in global politics, the impact of terrorism, and environmental issues. Presenting globalization in accessible language as a multifaceted process encompassing global, regional, and local aspects of social life, Manfred B. Steger looks at its causes and effects, examines whether it is a new phenomenon, and explores the question of whether, ultimately, globalization is a good or a bad thing.

www.oup.com/vsi